THE LIFE AND LETTERS OF
KATE GLEASON

THE LIFE AND LETTERS OF
KATE GLEASON

JANIS F. GLEASON

RIT PRESS

ROCHESTER, NEW YORK

RIT Press
90 Lomb Memorial Drive
Rochester, New York 14623-5604
http://carypress.rit.edu

Book and cover design by Marnie Soom
Printed in the U.S.A.

ISBN 978-1-933360-47-8 (Paper)
ISBN 978-1-933360-50-8 (Cloth)

Library of Congress Cataloging-in-Publication Data

Gleason, Janis F., 1934–
    The life and letters of Kate Gleason / by Janis F. Gleason.
        p.   cm.
    Includes bibliographical references and index.
    ISBN 978-1-933360-50-8 (cloth) — ISBN 978-1-933360-47-8 (pbk. : alk. paper)
    1. Gleason, Kate, 1865–1933. 2. Women engineers—United States—Biog-
raphy. 3. Engineers—United States—Biography. 4. Gleason Corporation—
History. I. Title.
    TA140.G54G54 2010
    620.0092—dc22
    [B]
                                                            2010029344

Contents

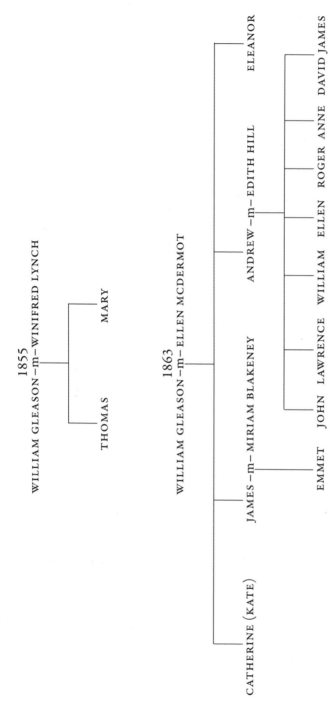

1855
WILLIAM GLEASON –m–WINIFRED LYNCH

THOMAS   MARY

1863
WILLIAM GLEASON –m– ELLEN MCDERMOT

CATHERINE (KATE)   JAMES –m– MIRIAM BLAKENEY   ANDREW –m– EDITH HILL   ELEANOR

EMMET   JOHN   LAWRENCE   WILLIAM   ELLEN   ROGER   ANNE   DAVID JAMES

1911
WILLIAM GLEASON –m– MARGARET PHELAN

Acknowledgments

M Y HUSBAND, JAMES S. GLEASON, grew up hearing his grandfather's negative version of Kate Gleason's contributions to the Gleason family enterprise, but it did not stop him from allowing me full access to the family files held at the Gleason Works in Rochester, New York. His encouragement and support made it possible for me to find the single most valuable source to my quest, without which this book could not have been written: the treasure trove of original material that was kept in the company's files. And without the organizational skills and meticulous devotion to history and fact of the late Nettie Bullis, I would never have had Kate's letters, and all of the other essential documents, from which to construct her biography. Nettie was James E. Gleason's right-hand woman at the Gleason Works, and she held many positions of importance in the company, including, at the time of her death, assistant secretary. Thank you, Nettie.

Many other Gleason employees have been interested in my project and have aided me in various invaluable ways, especially Don Allis, Valerie Barker, Marc-Henry Debard, Karen Essig, Ken Ferries, Elvira Hawkins, Gary Kimmet, Ginny Lalka, Marcia Smith, Dr. Hermann Stadtfeld, and Bruce Tyo.

Larry Rowland, son of Kate Gleason's personal secretary, and professor emeritus at the University of South Carolina at Beaufort, has been particularly helpful to me. He took my husband and me on a tour of sites that were pertinent to Kate's life and interests in Beaufort and put it all in historical perspective, and not least, shared the unpublished journals kept by his mother and Lillian Gilbreth of a European trip taken with Kate in the summer of 1924.

We spent several highly fruitful days with Ellen Gleason Boone in

South Carolina. Her memories of her "aunt Kate" and of managing the Beaufort property following Kate's death were quite clear and extensive. Ellen introduced us to people who had known Kate and she spent hours talking to us about the relationship between her father, Andrew Gleason, and his sister, and told us about a visit she had made with her mother to Kate's home in Septmonts, France.

I want to thank Jennifer Smith who visited Cornell University, her alma mater, on my behalf to research Kate's matriculation there and her relationship to President and Mrs. Andrew White and others. Also, Jennifer travelled to West Lafayette, Indiana, to the Purdue University Library, to explore the Frank and Lillian Gilbreth papers for references to Kate Gleason.

The East Rochester History Office, under the direction of the late Mary Connors, collected a great deal of material relating to Kate Gleason's enterprises in that community, and Mary was exceedingly constructive in arranging for interviews, sharing information about property owned by Kate, and recounting her place in the history of the village.

I am indebted to the Damas family of Septmonts, Aisne, France, who were so hospitable and informative; to Shirley Sponholtz, editor of *Old Time Trucks*, who cheerfully shared important knowledge about the early trucking industry; to Laurie Barnum and Deborah Hughes of the Susan B. Anthony House, who hung out the "Welcome" sign for me there; to Isabel Kaplan at the Carlson Science and Engineering Library at the University of Rochester; and to Mary Huth, Assistant Director of Rare Books and Special Collections at Rush Rhees Library at the University of Rochester, who was helpful on several fronts.

I am grateful to Janet Jimenez and Sandi Agostinelli for arranging an interview with their mother, the widow of Anthony J. Agostinelli, in early 1999. Mrs. Agostinelli's memories of her life in Beaufort, as a new bride whose husband was working for Kate, were rich and perceptive, and her sense of humor was in fine form.

I am most appreciative to a number of other people who granted me an interview: my father-in-law, Lawrence C. Gleason, who was one of Kate's nephews; Alice Wynd, who grew up across the street from Clones, Kate's home on East Avenue in Pittsford, New York; Mary Burrill, a great friend of Kate's sister, Eleanor Gleason; Richard Walter, whose parents had worked for Kate in East Rochester; Dreka Stokes, whose parents ran the Gold Eagle Tavern in Beaufort; and Margaret Sheper, whose husband was cashier of the People's State Bank in Beaufort.

Joseph W. Campbell, executive director of the Rochester Engineering

Society, Inc., gave me copies of the minutes from the meeting during which Kate was elected to membership in that organization, and shared some observations that did not appear in the minutes. Betsy Brayer, biographer of George Eastman, shared with me some letters that had been written by Kate Gleason to George Eastman that she had come across in the course of her research. Carrie Remis and Kathy Urbanic found records of "Kittie" Gleason's attendance at Nazareth Academy. Cousin Padraic Neville was kind enough to send me his account of a visit that he and his sisters, Jeane and Harriet, made to Tipperary in 1982 to research the Cleary family history. I turned to my good friend Dr. Edward Atwater for information about medical history questions I had from time to time. Edward was one of two people who read my first draft and made useful suggestions.

The other person who read that first draft was Karl Kabelac, a friend and mentor who taught me much about the rudiments of researching and the need to get as close to the source as possible. Karl has been my guide-star through the many years I have been working on this project—he has advised, suggested, encouraged and contributed every step of the way.

The enthusiasm with which David Pankow, director of the RIT Press, embraced this biography of Kate Gleason was a much appreciated affirmation of my own conviction that her story should be told. David's staff, especially Molly Cort, Marnie Soom, Patricia Cost, and Laura DiPonzio Heise, were all exceedingly helpful and patient with this neophyte to their industry.

And finally, I wish to thank my niece, Lisa Fitzpatrick, who has given me the benefit of her experience in the publishing business, and who introduced me to my personal editor, Susan Wels. I cannot imagine a more congenial or productive relationship than that which I have had with Susan, and I am eternally grateful to her.

Preface

KATE GLEASON SUCCEEDED in doing what no other American woman had ever attempted. A pioneer in engineering education, machine tool manufacture, banking, and low-cost residential development, she lived by a Latin motto, *possum volo*—"I can, if I will." She also lived by the words of Susan B. Anthony, her mother's great friend: "Any advertising is good. Get praise if possible, blame if you have to. But never stop being talked about."[1]

Periodicals and newspapers celebrated Kate's exploits throughout her life—calling her a "female mechanical genius" with a Midas touch, and she did nothing to discourage their attention. On the contrary, she believed that telling the world what a woman could do might inspire others to follow her example. Not all Kate Gleason legends, however, were based in truth. None other than Henry Ford credited her with inventing the Gleason gear planer, calling it "the most remarkable machine work ever done by a woman."[2] But as Kate herself told the *New York Times* in 1910, setting the record straight, "the nearest I have come to designing it is in having a father and brother smart enough to do it."[3]

What Kate Gleason did achieve in her singular lifetime, however, was itself remarkable. She was the first woman admitted to the engineering program of Cornell University. Later, as secretary-treasurer of the Gleason Works, her family's machine tool company, she was the only woman to hold such an elevated post in U.S. manufacturing. She was also the first woman member of the American Society of Mechanical Engineers and the German Engineering Society (Verein Deutscher Ingenieure). After retiring from the Gleason Works, she was the first American woman ever elected president of a national bank who had no family ties to the institution.

Kate didn't stop there. She bought tracts of land in East Rochester, N.Y.

and built factories and affordable housing for local workers, using innovative methods of concrete construction. She built a country club and a golf course, ran a trailer car company, and designed what was perhaps the first mobile camper pulled by an automobile. She bought land in Sausalito, California, and built houses there, too—and she constructed a hotel, a resort, and causeways on property she acquired in South Carolina. After the First World War, Kate became an angel to her adopted village of Septmonts in northern France, purchasing property and developing industry to rebuild the structure and spirit of the war-decimated town.

Kate Gleason, above all, was her own best invention. She was a brave, hard-driving, sometimes manipulative businesswoman and dreamer who thought all things were possible for the determined. Raised in the avant-garde of late nineteenth-century feminism, she believed herself capable of achievements far beyond a woman's traditional role. Kate relished her exceptionalism, and she often used it to her advantage. In the early days, she remembered,

> I was a freak; I talked of gears when a woman was not supposed to know what a gear was. It did me much good. For, no matter how much men disapproved of me, they were at least interested in seeing me, one distinct advantage I had over the ordinary salesman. I dealt wholly with men—no women were then running factories and foundries."[4]

However, while Kate followed a male path, her approach was feminine. She was communicative and emotionally sensitive—useful qualities for any salesperson—and promoted the products she was selling as well as herself.

Many people took note of her unconventional, uncompromising career. Susan B. Anthony saluted her as "the ideal business woman of whom I dreamed fifty years ago."[5] Playwright Eugene O'Neill, too, saw her as a remarkable example of female power—although, perhaps, not in ways that Kate would have appreciated. She was the inspiration for a play, part of a cycle he never completed, called *The Life of Bessie Bowen*—about a squat, greedy, unattractive woman who enslaved men in her drive for business success. Kate Gleason had plenty of faults, but not the ones O'Neill ascribed to Bessie Bowen. Kate's figure was, in truth, rather squat, and she fought the battle of the bulge most of her life. But greed was never Kate's motivation, and she was attractive, generous, and helpful to the men around her. Although she became prominent and powerful, she never lost her principled code of ethics and conduct—nor, for that matter, her good humor and kindness. Unable to complete the cycle, Eugene O'Neill burned his manu-script and notes in 1952, nineteen years after Kate Gleason's death. Despite her taste for notoriety, I am certain she would have been relieved.

Kate would, however, have been pleased by another tribute to her life's achievement. In 1998, Rochester Institute of Technology named its engineer-ing school the Kate Gleason College of Engineering—the first to be named after a woman engineer and a beacon for those inspired by her example.

Hers is a life story worth telling. However, anyone who undertakes to write about the life of someone else—especially someone who is deceased and no longer able to defend herself—is assuming a delicate responsibility, since the record is never complete and therefore unavoidably flawed. My aim, whenever possible, is to let Kate tell her story in her own words, so far as I am able.

I first encountered Kate's life and legend when I married into the Glea-son family in 1959 and moved to my husband's home in Rochester, New York. As a Californian, it was the first time I had been east of the Missis-sippi River. During those initial weeks in Rochester, I was trotted around to pay calls on various relatives. One of the most interesting visits was to the home of my husband's great aunt Eleanor, Kate's younger sister, in Pittsford, New York. Kate had built the house in the first decade of the twentieth century. It was designed in the style of the Alhambra in Granada, Spain, though on a lesser scale, and Kate had named it Clones, after her mother's birthplace in Ireland.[6] By 1959 it was still very exotic, though a bit seedy around the edges, and it was all the more fascinating to me because of the stories I had been hearing about Kate Gleason. Those tales, exchanged at family gatherings, were delivered with a disapproving and contemptuous tone; they painted a picture of a spoiled, bossy, overly confident, eccentric woman who took credit for the accomplishments of others.

But as I became more integrated into Rochester life, and met more and more people, I began hearing quite a different story about Kate Gleason. I learned about her many remarkable "firsts" and her good humor, generosity, kindnesses, intelligence, and charisma. From time to time, an article would appear about her in some local publication, such as *Upstate Magazine* or the *Brighton Pittsford Post*. The writers were always complimentary to Kate, and they often expressed outright admiration. People also made comments to me, on occasion, about a story they had heard or an experience they had had with Kate, and I gradually became aware of a discrepancy between the family's view and opinions of her outside the Gleason circle.

Shortly before Eleanor died, she gave my family two scrapbooks that had belonged to Kate. One of the books contained photos and drawings of Clones and the other pictures and memorabilia from Kate's property in Septmonts, France. Eleanor also told us that she had been left Kate's personal effects, including stacks of papers, letters, and manuscripts, but that she had burned them. Although she gave no explanation, her manner suggested that we were all better off without such foolishness around. Eleanor may have had some reasonable motive for committing Kate's history to ashes, but I rued the loss of that material, and Kate's story continued to intrigue me.

Since my husband was working at the family's machine tool company, I asked him if any information about Kate could be found there.[7] That was when the pieces of her story began falling in place. Several filing cabinets on the top floor of the Gleason Works contained personal family papers and mementos, and I began snooping in earnest. There were photographs, newspaper and magazine articles, a black leather-bound book of Kate's jottings over a long period of years, and letters—perhaps a hundred, most written by Kate to her brother Jim when he was a student at Cornell University and she was working in the Gleason factory in Rochester, and some written to Kate by her father. Many of the letters are grammatically incorrect and contain misspellings, especially those written by William Gleason, and I have left them all as they were written, without corrections. Once I read all of those letters, I knew that the family's assessment of Kate was badly askew.

With the perspective of an outsider and the position of an insider, I will attempt to give some insight into this remarkable woman. I have no hidden agenda, no axe to grind, no point to make. This will be as straightforward a history as I can make it.

Jan Gleason
April 2010

NOTES

1   Bennett, "Kate Gleason's Adventures," 170.
2   Ibid., 43.
3   Kate Gleason, letter to the editor, *New York Times*, May 21, 1910, 8.
4   Bennett, "Kate Gleason's Adventures,"170.
5   Susan B. Anthony, inscription to Kate Gleason in vol. 1 of *History of Woman Suffrage*.
6   Pronounced Cloh-ness and sometimes spelled Clonos, other times Clones, in Kate's letters and papers.
7   James S. Gleason, C.E.O. of the Gleason Corporation from 1981 until 2002, is currently chairman of the company.

Foreword

I JUST LOVE this account of a woman who took charge of her own un-conventional destiny.

To find excitement in doing work that is productive in today's world is what most of us would like to do. Kate Gleason achieved that in a non-traditional field. She has earned more of our respect, 100 years or more later, than she probably had received in the intervening years. It would be nice to say that it is because we no longer face the same circumstances that challenged Kate's credibility, but in fact, her life still holds valuable lessons for all of us.

Today, engineering students and their prospective employers are less likely to use gender to determine aptitude and talent, though statistics prove that workforce barriers persist. Despite diversity-promoting campaigns to encourage young women to pursue this field, the mechanical engineering community of which Kate Gleason was a part remains a male-dominated discipline. The American Society of Mechanical Engineers (ASME) was part of that professional network for Gleason. I, as president of ASME some 70 years later (1986–87) and as the first woman president of a major engineering society, applaud the example her life sets for students today and her adventurous, risk-taking, can-and-will-do spirit.

Some 20 years after Kate's death, I began a career as a mechanical engineer that spanned nearly 40 years at the General Electric Co. (GE) in Schenectady, New York. I retired from GE's Global Research, formerly known as Corporate R&D. My area of expertise was heat transfer and fluid flow, which I applied to gas turbines, nuclear energy and space satellites, among other fields. R&D put me amid rapid technological changes that were to be significant in our future. That's where I wanted to be. I was one of the first engineers to work on the heat transfer of nuclear reactor cores and on the first satellites designed by GE. My areas of expertise never changed, just my projects. I had planned to work five years or as long as it was fun. Those five years stretched into thirty-eight. I never seemed to get bored with my work. I tend to see a common thread in Gleason's later years after leaving Gleason Works. She identified various projects that not only challenged her, but had an element of civic responsibility as well. They stretched her capabilities and fed her spirit.

Just as Gleason found professional support, I too found encouragement within the engineering community, not only at GE but also at ASME. For Gleason, her friend Henry Sharpe (of Brown & Sharpe) helped open doors for her throughout Europe. And most likely because of their friendship, Lillian Gilbreth followed Gleason into ASME membership in 1926, becoming one of the most celebrated women of ASME. Lillian, along with her husband, developed time and motion studies that improved industrial efficiencies. Those studies were among the many historically significant contributions by management leaders within ASME membership. Chapter 11 of this book, in particular, hints at many prominent industry relationships overlapping with Gleason's social life as Herbert Hoover entered the White House. After reading this book, I can see she epitomized the kind of leadership ASME would have sought among its membership, though I'd venture that ASME's appeal to her included her desire to be the first woman member. Her zeal would have been useful in offsetting any resistance that she might have met.

To understand what ASME was like then, during Gleason's years of participation in ASME activities from 1914 through 1933, consider that ASME celebrated its fiftieth anniversary (1930) as a 20,000-member (currently 120,000-member) professional society, built by industry leaders who were well-traveled, focused on active connections and interested in projecting a national voice for the profession. ASME's influence flourished through standards development and research initiatives. Gleason's nearest well-known contemporaries were Thomas Edison (1847–1931) and Henry Ford (1863–1947), both ASME members. George Westinghouse (1846–1914), who had been ASME president in 1910–11, had left his impressive industrial legacy in an age when corporate research dominated technological advancements.

The tenor of the 1920s was captured in what historian Lewis Mumford called the "Neotechnic Age" of abundant electrical power, and which ASME historian Bruce Sinclair describes as the "Machine Civilization."

Managerial efficiencies, high-speed steel advances, and innovations in commercial transportation and mass production led to a technologically advanced society that outpaced social acceptance to some degree. In that climate, Gleason's visions of commercial opportunity were insightful and well executed. Kate's later philanthropic efforts, including affordable housing, mirrored the engineering ideal of working for the benefit of humanity.

Kate Gleason's optimism, her willingness to learn her business from the bottom to the top, and the realities of balancing business and social concerns are all part of the voice you will hear in this book. She was a woman who enjoyed being first in her field. She was a woman who learned how to work in a male-dominated profession and create a network of her own friends and associates around the world. Her father's direct influence is discussed more herein, but I'm impressed with her mother's suffragist friends and how their strengths certainly helped shape Gleason's future. She was a very young woman of 18 or so when she set out to be a sales representative for her father's machine-tool business. She took it global. Then, when it was time to move on, I think that same high-spirited nature led her to ever greater achievements that enriched her life and the lives of others, through the changes she made in the world.

This biographical record is a remarkable treasure. Kate lived in a time of hand-written letters and "wires" as a means of communication. Reading them gives one the wonderful feeling of being a part of Kate's life as it transpired. One of the many things I appreciate about the book is that the letters give us amazing insight into the Gleason family and their interaction within the business world and society at large. Through her own voice, the letters immerse us in her personal story, intriguing us wholly in a factual but not dispassionate way. It was an age that was archived through letters and diaries, with an intimate glimpse into people's lives, in a way that we may not see again in this world of email and digital text messaging.

Jan's book gives great insight into the personal life surrounding this most amazing woman who went after what she wanted with integrity and perseverance. Today's engineer absolutely needs more examples of entrepreneurial drive, and the example that Kate Gleason sets is as relevant today as it was remarkable then. The breadth of her interests and unflinching pursuit into many diverse business ventures, both during and after her work at Gleason Works, impresses me both for her energetic managerial expertise and technical knowledge.

The moral of this story is that if you have a passion for and are dedicated to your work, and if you want to take charge of your own destiny, let

Kate be your mentor. If you've wondered who blazed the trails in the early 20th century generation, read this book to touch upon additional true stories that unfold therein. And if you want a glimpse into history through the eyes of an industrial entrepreneur, you will find it here. It is a "good read."

Whether you are a man or woman, please do note the importance of a mentor and do break with tradition to inspire and nurture true talent found in others.

Nancy Deloye Fitzroy, D.Eng., D.Sc., P.E.
ASME President 1986–87

PART ONE
BUILDING THE FAMILY BUSINESS

CHAPTER I

The Early Years

IN THE FIRST YEARS OF THE TWENTIETH CENTURY, Kate Gleason built herself a romantic Moorish palace in sedate Rochester, New York. Inspired by the lavish description of the Alhambra in Edward Bulwer-Lytton's nineteenth-century novel, *Leila, or The Siege of Granada*, the mansion was a preposterous flight of fancy, and the kind of bold stroke that naturally appealed to Kate. It proclaimed her independence and consequence, and, by extension, the possibilities that were open to any woman with the ability and courage to look beyond banal customs of the day.

Kate's daring Alhambra reflected her iconoclastic, visionary side, and perhaps a wishful reimagining of her own origins. Her father's mother had been a Cleary with ancient ties, in the misty Irish past, to a Norman castle complete with drawbridge and portcullis in Tipperary. Kate named her modern aerie Clones, after her own mother's birthplace to the north, near Belfast. If the McDermotts, on her mother's side, and the Cleary family, on her father's, had ever known power and wealth, however, it had crumbled over many generations. Kate's roots on both sides were planted in hard soil, among the struggling poor.

In 1836, four years after her grandmother, Mary Cleary, married Thomas Gleason, their future was so bleak that Thomas left his wife and two baby sons in Tipperary and boarded the ship *Godspeed*, bound for Quebec, seeking a better life for his hungry family. Mary had been forty-two and a childless widow when she wed Thomas. Now, with two-year-old James and newborn William to feed, she soon found herself a widow for the second time. Months after Thomas arrived in Canada, he secured employment as a local militiaman and was killed on riot duty in Montreal. Mary and her babies moved to a nearby farm owned by her brothers, and for the next

1

twelve years they lived in a two-room cabin, surrounded by peat bogs that yielded fuel but little in the way of food. Mary, William remembered, gave her young sons pieces of peat to cut up on their plates so they would know how to use a knife to cut meat, if they ever got any. Although William's children, in later years, were skeptical about the tale, the Gleason family, without a doubt, lacked enough to eat, especially once the potato famine swept Ireland in 1845. The potato blight destroyed the country's staple food crop, triggering a human and ecological disaster, since a third of Ireland's people relied almost entirely on potatoes for food. In Tipperary, where mortality rates doubled from 1846 to 1847, the death toll from starvation was among the highest in the country.

Mary Gleason had never been further from home "than her own feet or a donkey cart could carry her,"[1] but she was soon desperate enough to contemplate the dangerous voyage across the ocean to North America. Nearly a million and a half people fled Ireland—80 percent of them to the United States—between 1845 and 1851, many perishing of hunger and disease in dank coffin ships on the Atlantic crossing. In 1848, Mary, now fifty-eight, with her two sons—fourteen-year-old James and twelve-year-old William—left Ireland aboard the *Borneo* on the dangerous six-week passage across the Atlantic. On a bitterly cold day in November, the family at last disembarked in New York City, and soon afterwards made the 230-mile journey north to Oswego, New York, where the Gleasons had family. From there, Mary and her sons crossed Lake Ontario by tugboat to Canada and their final destination, Kingston, Ontario. During their frigid night crossing, the last passage before navigation shut down for the winter, Mary and the boys huddled on an open deck under a comforter, one of the very few possessions she had brought from Ireland.

In Kingston, at the head of the St. Lawrence River, young William was able to find employment in a grocery store. Years later, he remembered filling standard orders for farmers who came in once a week to re-provision: a pound of sugar, a quarter pound of tea, and a fifty-cent gallon jug of whiskey, with a corncob stopper. The boys were ambitious, but there were few opportunities in Kingston. So in 1851—after family friends, the Connells, moved to Rochester, New York, and reported that the town had promise—Mary, James and William moved to Rochester, some one hundred miles southwest, to start again.

William quickly found employment on a farm, where he helped pick berries in the morning and took them to market in the afternoon, but he knew there was little future for him in that job. Another opportunity luckily, came his way thanks to his friend John Connell, who worked in a machine shop as an apprentice. William was attracted to that line of work and soon found a job as a night helper in a machine shop on South Avenue. One evening, when a rocker-pin failed in one of the shop's engines, William took it upon himself to machine a new one, a task that required training and expertise. In the morning, the foreman scolded William for exceeding his responsibilities, but the owner overheard the conversation, admired his initiative, and asked him to start reporting for day work as an apprentice-in-training. This was the chance William had been waiting for, and it gave him the foundation for his life's work.

Soon, with a promising job, William, at age nineteen, married Winifred Lynch; a year later, in 1856, their first child, Thomas Francis Gleason, was born in Rochester. By 1857, however, the economy was weakening, and jobs were scarce. Even though William suffered badly from seasickness, he relocated his family to Chicago, where he took a better-paying job as a stoker on a lake steamer. His daughter Mary, born in that city in 1858, died the following year, and Winifred succumbed to tuberculosis shortly thereafter.

William, now a widower, returned to Rochester with six-year-old Tom. His mother, Mary, supervised the boy while William worked in a machine shop at the corner of Mill and Furnace Streets and took night school classes in mathematics. His older brother, James, was working for the army as a locomotive engineer. In 1863, at the height of the Civil War, James was killed in Tennessee when Confederate soldiers placed obstructions on the tracks, derailing his train. William now became the sole supporter of the Gleason family. He was needed at home, and therefore exempt from service, but he was intent on aiding his adopted country and enlisted anyway. To his family's relief, the company's quota was filled before William's name came up, so he remained in Rochester.

On September 11, 1863, at age twenty-seven, he took a new wife, nineteen-year-old Ellen McDermott. She and her family had emigrated to Rochester from County Fermanagh, near Belfast, Ireland, in the same year the Gleasons came to America. When Ellen was ten years old, she had worked fourteen hours a day in a Rochester cotton mill to help feed her family. Soon after they married, William left Rochester for Hartford, Connecticut, where he had heard that Colt's Armory, a gun manufacturer, was hiring machinists. He took with him just enough cash to pay for a week's board in advance, leaving the rest with his family, and went to Colt's, only to find that no jobs were available after all. William was never one to give up easily, and he hung around the Armory until he noticed a lathe standing idle on the factory floor. Putting on his overalls, with little to lose, he strode over to the lathe and got to work. By the time the other employees noticed him, he had produced a prodigious amount of high-quality material. The foreman, impressed by his workmanship and amused by his enterprising spirit, promptly hired him as a mechanic.

The Pratt and Whitney Company was also in the gun manufacturing business in Hartford, and William worked for both firms while he lived in Connecticut. It was a time when craftsmen were developing the mathematical precision necessary for mass production, and many men the firms employed were later prominent in the emerging machine age.[2] By 1865, the Civil War had come to a tumultuous close, and William, his talents honed in Hartford, decided to return to Rochester and strike out on his own. With his friend John Connell, he opened a small machine shop at the corner of Furnace and Mill Streets, in a neighborhood known as Brown's Race. The area, near the upper falls of the Genesee River, had plenty of waterpower and attracted a considerable portion of the city's industry. William's new firm, called Connell & Gleason—the forerunner of today's Gleason Corporation—manufactured engine lathes, planers, and woodworking machinery.

That same year, on November 24, 1865, William's wife, Ellen, gave birth to a daughter. The Gleasons named her Catherine Anselm, after Ellen's mother, Catherine McDermott, but they called their healthy, ruddy baby girl "Kittie" and later "Kate."

At the time of Kate Gleason's birth, Rochester was called the Flower City because of the prominence of its nursery and seed businesses. With its factories and enterprising merchants, the city boasted fifty thousand residents, 35 percent of whom were immigrants, mainly from Ireland and Germany.

The Gleason family settled in the city's Second Ward in 1865 and 1866, at 105 Oak Street. It was a large household consisting of William and Ellen; young Tom and Kate; William's mother, Mary Gleason; and Ellen's parents, Luke and Catherine McDermott. The next year the McDermotts relocated to Platt Street, and the Gleasons moved to the east side of the Genesee River, settling in an area known as "Little Dublin" because of its large Irish-born population.

William's business was prospering, thanks in part to his ingenious approaches. He earned his first patent on November 12, 1867, for a device he designed, called Gleason's Patent Tool Rest, which held tools in a metal working lathe. In 1868, he and John Connell brought in a third partner, a

Scotsman named James S. Graham. They called their growing enterprise Connell, Gleason and Graham, and hired three workers to boost their production of engine lathes, planers, and woodworking machinery.

As the business grew, so did the Gleason family. On November 25 that year, one day after Kate's third birthday, Ellen delivered a son, James Emmet Gleason. Four years later, in 1872, the expanding family moved to 104 Platt Street, by the corner of Jones Street, in an industrial neighborhood near the upper falls of the Genesee River. Another son, Andrew Chase Gleason, was born on November 16 of that year.

Since the early 1800s, western New York had been a hotbed of religious and reform movements, from abolition to utopian experiments and women's rights. Feminist Susan B. Anthony lived in Rochester, and Ellen Gleason, a passionate advocate for women's voting rights, counted Anthony and many suffragists among her friends. Kate later attributed her success in life to Anthony's inspiration. Just two weeks before Andrew was born, in November 1872, the suffragist, along with forty-nine other women, was arrested for registering to vote and daring to cast a ballot. Anthony was subsequently indicted, convicted, and fined one hundred dollars for the transgression, which she refused to pay. One can imagine the anger and outrage expressed at 104 Platt Street. Except for the birth of her son, it was a painful year for Ellen. She lost her mother, father, and twenty-one-year-old brother, probably to influenza, and her seventeen-year-old sister died in February 1873.

William, meanwhile, had been having disagreements with his partners at Connell, Gleason and Graham about the future direction that the firm should take. William was convinced that the expanding railroad system would increase the demand for metal-working tools, but his partners wanted to specialize in woodworking equipment. As a result of these differences, William left the firm and joined Kidd Iron Works, one of the largest machine shops and foundries in western New York, as a partner and machine shop superintendent.

Kidd manufactured lathes, planers, upright drills, boring machines, steam locomotives, caloric engines, railroad car wheels, and iron tools. Its three-story building had a sixty-eight foot frontage on Brown's Race and ran back eighty feet to the high bank of the Genesee River. Next door was a cotton mill, constructed before 1815, whose bell called workers to their posts at five in the morning, summoning thirty men to Kidd Iron Works for a sixty-six-hour work week under harsh conditions.[3] The electric light had not yet been invented; uncovered gas flames illuminated machines, singeing the eyebrows of many machinists who tried to get close enough to the equipment to measure a set-up. Drawings were made on Manila paper that was fastened and shellacked to a wooden board. Men furnished their own pails for washing up, heating icy water in winter with hot slag from the foundry. Toilet facilities, according to one account, consisted of a perilous "four-holer" perched a hundred feet over the churning tailrace.

William was soon flourishing at Kidd. He continued to design new types of machinery and in October 1873 earned a patent for improved screw-cutting attachments for lathes. When the firm of John T. Noyes & Sons of Buffalo, New York, asked Kidd to develop a machine tool for use in milling equipment, William set to work on the design. His revolutionary new machine made it possible, for the first time, to accurately and efficiently mass-produce bevel gears, for transmitting power around a corner. That September, however, bank failures had triggered a deep depression, and Mr. Kidd decided that his firm could not afford to build the new device. William, undaunted, built the novel gear planer at his own expense and shipped it to John T. Noyes & Sons on October 27, 1874. His original machine operated continuously for forty-five years and revolutionized the production and use of bevel gears, opening vast new possibilities for the transmission of motive power.

In 1875, William became the sole owner of Kidd Iron Works and changed its name to William Gleason, Tool Builder. In 1876, after receiving a patent for a gear-cutting and dressing machine, he took orders for six of the devices and five more the following year, despite the lingering depression. He was devoting all his attention and energy to the success of his shop; the well-being of his growing family depended on it, and he was hardworking by nature. The years ahead required no less of William Gleason.

Although the Gleasons were Irish Catholics, they were independent thinkers who rarely bowed to pressure or convention, especially when it came to their children's education. The Gleasons sent Kittie to Nazareth Academy, a Catholic girls' school near their home, but she was transferred to public school by the time she entered high school. Her brothers never attended parochial schools, and all of the children attended non-Catholic colleges. The Gleasons shared Susan B. Anthony's strong belief that the public schools of their day were best at assimilating new Americans. As millions of immigrants streamed into the United States, with their own languages and customs, public schools taught their sons and daughters the advantages and responsibilities of living in American society, easing their entry into the amalgam of American life.

The Gleasons' preference for public education, however, had a cost. In 1858, Bishop Bernard J. McQuaid had arrived in Rochester and energetically promoted the parochial schools, informing Catholic parents that they had a spiritual obligation to send their children to those institutions. Those who did not comply faced a severe penalty: denial of the sacraments.[4] William Gleason, nevertheless, stood his ground. The bishop hectored him, unsuccessfully, about enrolling his boys in Catholic schools, and as punishment for his intransigence, refused to allow him to take Communion. Dumbfounded and outraged by what he saw as dishonorable treatment, William never again set foot in a Catholic church. "He hated tyranny whether in church or state. His dominant characteristic," according to his son Jim, "was absolute independence of action and thought," and he was determined to make the best choices for his family.[5] Although Ellen and the children continued to attend church, they did so with varying degrees of devotion. Kate, for instance, was thought to be quite devout because she was often seen reading from a prayer book. Her sister later explained, however, that Kate routinely wrapped a prayer book cover around whatever novel she happened to be reading at the time; like her father, Kate had an independent nature, unbowed by authority.

Both her parents were broad-minded, and Kate grew up in an atmosphere that supported and encouraged her independence. Although her grandmother, Mary Gleason, ignored her and lavished attention on her brothers, Kate had a mind of her own from an early age and was determined to beat the neighborhood boys at their own games. "If we were jumping from the shed roofs," she later recalled, "I chose the highest spot; if we vaulted fences I picked the tallest. I was husky and able, and to this I added a bit of recklessness that carried me through. It took just that added bit of [derring-do] to outdo the rest."[6] Friends and neighbors, watching Kate's antics, used to say she should have been a boy. "They were justified," Kate said, "for I was trying my best to be as nearly a boy as I could." She felt keenly "that girls in this world were accorded second place, and I resented being second." She wore her hair short and straight in a day when girls wore long curls or braids, and "I played with the boys," she stated. "They didn't want me, but I earned my right."

Kate adored her older brother Tom, William's right-hand man, who shouldered many business responsibilities. In 1876, however, Tom, age twenty, died suddenly from typhoid fever. It was a devastating blow and a turning point in Kate's life and William's business. Eleven-year-old Kate idolized her father, who keenly missed the help and support of his eldest son. One day after school, soon after Tom's death, she overheard William say, "Oh, if Kate had only been a boy!" Kate hated failing her father in any way. "Although I was young," she remembered, "I was a great big girl and looked much older," and after Tom died, Ellen thought it would be wise for her daughter to learn something about the business. So the next Saturday, Kate "walked down to the shop, mounted a stool and demanded work. Father smiled and gave me some bills to make out. From that time on," she added, "I worked regularly."[7]

NOTES

1   Eleanor Gleason, speech, Jan. 29, 1931.

2   During that era, Pratt and Whitney employed Worcester Warner and Ambrose Swasey of the Warner and Swasey Company; George Bardons of Bardons and Oliver; A. F. Foote of Foote, Burt and Company; E. C. Henn and Reinholdt Hakewessel of the National-Acme Manufacturing Company; Frederick Gardner of the Gardner Machine Company; J. N. La Pointe of J. N. La Pointe Company; Frederick Geier of Cincinnati Milacron; Edward Payson Bullard of Bridgeport Machine Tool Works; and William Gleason. The latter three gentlemen all worked at both Pratt and Whitney and Colt's Armory. The Brown and Sharpe Manufacturing Company of Providence, R.I. was also in the vanguard of manufacturing parts that were precise enough to be interchangeable. One of the men who trained in this Providence shop, Henry M. Leland, went on to found the Cadillac Car Company, and later, the Lincoln Motor Company. Mr. Leland was considered to be the "Master of Precision" and was the first executive in the U.S. automotive industry to insist on absolute accuracy of the parts used in his products.

3   Much later, after a fire in the cotton mill, Kate's brother Jim bought the bell and installed it in the Gleason factory powerhouse on University Avenue in Rochester. It is possible that their Irish immigrant mother, Ellen McDermott Gleason, had been summoned to work in the cotton mill by that same bell, which is now mounted at the entrance to the Gleason Works as a reminder of Rochester's long labor and industrial history.

4   McNamara, *Diocese of Rochester*, 166.

5   Roy Rutherford, "James E. Gleason Acclaimed as First Citizen of Rochester; Mechanical Genius Heads Plant Which 'Gears Up World'," *Rochester Democrat and Chronicle*, October 7, 1945, B6.

6   Bennett, "Kate Gleason's Adventures," 168.

7   Ibid.

CHAPTER 2

Sibley Sue

ALTHOUGH WILLIAM SOLD FIVE GEAR PLANERS IN 1876 and six in 1877, the banking crisis hit Rochester hard, and by 1878 he was hanging on by sheer force of will. Industry and investment were at a virtual standstill. That year the firm sold only a single planer, and the following year sales were limited to three machines. The collapse of business was a searing experience for William. "To his last days," his daughter Eleanor recalled, "the man at the head of that struggling business had every so often a night-mare in which he dreamt that it was Friday night and he had no money to meet Saturday's payroll. Perhaps it was a certain seriousness brought on by those worries and responsibilities that caused him to be called sometimes 'The Governor,' more often 'The Old Man,' even before he was fifty."[1]

In 1879, William cut his work force from a hundred fifty men to three; he had his back against the wall. "We were working only five hours a day," an employee remembered, "and on Saturdays [we] carried home more money than he took home to his family; he always managed to scrape up money for us."[2]

To keep costs down, and since the firm employed just a handful of workers, Kate, at age fourteen, decided that she could handle the book-keeping by herself. "Father backed me up then," she said, "as he backed me up all of his life," and he gave her one dollar, her first pay, after her first day at her new job.[3] Somehow, Kate lost the dollar on the way home, to the consternation of her mother and grandmother, but it was never the pay that motivated Kate; it was the challenge of the business and her drive to help her father make it successful. Soon she was spending many hours a day at the firm, and she had no time for more feminine pursuits. Every day, she recalled,

I got up at four a.m. and studied, went to school at eight, got out at one, went home to dinner, then to the office, worked until six, went home, ate supper, and went to bed. The neighbors used to expostulate with Mother, and when she wanted me to go to dancing school, I rebelled. She told me Jim, my younger brother, was so shy that I must go to stir him up a bit; I found later that she had told Jim I was so uncouth I must be made more civilized, and so we each endured it for the other, both hating it with all our hearts.[4]

Soon the Gleason firm began prospering again, thanks in part to the boom in the oil business. From 1878 to 1880, Standard Oil was constructing the largest pipeline in the world in the Bradford oil fields, in western Pennsylvania and southwestern New York. The Gleason business, with its reputation for fine machinists' tools, had designed lathe features especially for the oil market. William's customers were loyal. In 1881, his business had recovered enough to allow him to buy the firm's building for $25,000, and he and Ellen celebrated the birth of another daughter, Ellen (who later called herself Eleanor) on March 4th; their family was complete.

Kate, now sixteen, was already graduating from high school at the Rochester Free Academy, the jewel of Rochester's public school system.[5] The academy offered a first-rate education, and many young Catholics sought admission there, despite the strong objections of Bishop McQuaid. Most academy students ended their formal education at graduation, but Kate had other ideas; she wanted to go to college.

She set her eye on Cornell University in nearby Ithaca, New York, the first major coeducational university. Cornell had admitted its first woman student in 1872, four years after its founding, and on September 16, 1884, Kate took its entrance exams in English grammar, geography, physiology, arithmetic, plane geometry, algebra, and most likely French and German.[6] Although mechanical engineering was an exclusively male field at the time, she passed her exams and entered the four-year Bachelor of Engineering program at Cornell's Sibley College of Engineering and Mechanic Arts, founded in 1870. It was a newsworthy event. "Miss Kate Gleason of Rochester, N. Y.," the *San Francisco Bulletin* reported, "is studying practical mechanics in Cornell University and is the only lady student in that department."[7] Kate was the first woman to study mechanical engineering at Cornell and the first of the school's pioneering "Sibley Sues."[8]

She joined male engineering students for shop work, and her other university courses included German, geometry and conic sections, and free-hand drawing. Kate was having an exciting time, but she was missed at home. On September 24, 1884, her brother Jim wrote her from Rochester,

> Ma has not got over mourning yet,—I think she will be down to visit you in about a month, although no definite time is fixed. She will then bring you all you asked for in your letter, together with the writing machine, for her own benefit as she can not decipher your letters in less than half a day as they are written so fine and closely.
>
> In looking over the paper the other evening, I discovered that I had passed the entrance examination to the Free Academy ... Andrew took the prize today at school for the best composition, which was written without any aid from anybody during school hours. Business is as bad as ever at the shop. The last time Pa went out for orders he procured one but had three or four promises ...
>
> Your affectionate bro,
> James Gleason
>
> P.S. You had better join that quoir you spoke of, and sing at High Mass.[9]

A letter from her father, written on November 11, was less discouraging about the business:

> I am glad to hear that you like your studys and am anxiously waiting for the time when you will be assisting me. Mr. Burghardt and myself don't find any trouble in running the shop. There is thirty-five men to work and at present have all they can do ... I think I am holding my own. If I do I am satisfied as the worst is over. The men have got used to the smaller wages ... Jim is working hard and while I have had a share of the hardships of life, having to deal with bad and willful children is not any part off them but on the other hand can be classed with the comforts of life and makes it worth living.
>
> Yours truly,
> Wm. Gleason[10]

Kate, like all women students at Cornell, lived in Sage College, a residence named for Cornell benefactor Henry Sage. Sage had given $250,000 to Cornell, provided that "instruction shall be afforded to young women ... as broad and as thorough as that now afforded to young men." All women, he believed, "should have the liberty to learn what they can, and to do what they have the power to do."[11] Everyday freedoms, however, were limited at Sage College. Women lived under the watchful eye of a female matron known as the "Warden," who enforced strict rules restricting their activities and social behavior. Women students resented the idea that they could not be trusted to behave properly without an overseer, but the voices of the "inmates" were too few to be heard. In a letter to Jim, Kate complained that the Sage rules were an insult to the girls. When the fall term ended on December 19th, she went home to Rochester and enjoyed her liberty over the holidays, returning to Cornell in early January for the winter term.

While Kate was at college, William had hired a replacement for her at the generous wage of eleven hundred dollars a year, evidence of the value that he placed on his daughter's abilities. In a letter dated February 1, 1885, he told her that "I had a hard month in Jan. to meet my payments but have got through all right ... I am going to have a daily paper sent to you as I don't want you to loose track off the modern world while clearing away the debris off old Greece and Rome for ancient information."[12]

There was little danger, however, that Kate would be distracted by the classics. Her winter term courses were German, algebra, free-hand drawing, instrumental drawing, and more shop work, and in the spring she studied German, trigonometry, descriptive geometry, text, drawing, and more shopwork. Her classes must have been challenging, because William wrote reassuringly to her on April 10: "I can tell you there is other things to make a person feel blue besides a 1/2 per cent discount in a class average standing."[13]

He needed reassurance, too. Business conditions remained dismal, and William and his customers were faced with discouraging financial obstacles. In order to sell their products, they often had to accept I.O.U.s from buyers, and then barter those I.O.U.s when they needed to make a purchase or provide banks with collateral. Notes passed back and forth between companies, many of them worthless, and businesses exchanged machines and other equipment instead of cash.

On May 10, 1885, William wrote Kate explaining his financial troubles and encouraging her to drop out of Cornell and come home to help:

> I have just returned from the sale of tools caused by the failure of the Miller Co. of Canton, O ... I don't think I will loose more than one or two hundred dollars ... Now for the heavey part. I suppose I might as well tell you at once that Prentiss is busted and I am left to the tune of $3500.00. When I wrote to you last time I told you that I had been to New York and by takeing machinery I reduced it to $2500.00. I counted a few days ahead. There was two notes due on the 16th of April which he assured me that he would

pay and as it was then within a few days off the time and as he seemed so confident of paying them that I commenced to figure the amt. he owed me as being $2500.00. Sometimes it seems to paralyse me and I feel as if it is only a dream but the stern reality is ever present. You have not any conception of the feeling that it has produced in me. It is one that I feel as if I am ashamed to tell of it and have not said mutch about it ... As might be supposed I am close pressed for money and have to be concocting schemes all the time. Add to the above the fact that prices are so low that there is very little profit. On reading your letter to Ellen I see that you feel it in your bones and that you have decided to stay at home and help the comeing year. I am glad you feel that way and as you say the college will not stop. You can return in another year. I think I am going to keep Jim home also. I will then feel that I have help and with the plant of tools that I have and three of us giving our undivided attention to the business it wont be long until everything will be bright again.[14]

Kate must have readily agreed to leave school, and on May 14 William wrote,

> Dear Daughter,
> Your kind letter is received and in reply I will say that it is my wish to have you stay there until the end of the present term ... With you and Jim to work I am sure we will make up this year what shrinkage and loss toock out last year. I am going as soon as you are home to give over the office business correspondence and all to you and I will put Mr. Burghardt to running the gear cutters. I am going to run the shop myself and with the three of us to work I will feel more encouraged and am confident that one year will place me all right when you can resume your studies.[15]

Kate adored her father and did not hesitate to do as he asked. Still, leaving Cornell was a wrenching prospect, "my first big sorrow," she recalled:

> My heart broke utterly. I took Father's letter out on the campus and sat under a tree where I thought no one would find me, and wept and wept. I had planned to finish the engineering course. I was the only woman in it, and it meant so much. As I sat there one of my friends who had been a pal saw me and came to me. He asked what was the matter. When I sobbed that I had to leave, he choked up and said brokenly that he was awfully sorry, but that just at present he couldn't be more than a brother to me. My tears stopped. I tried to convince him that I was crying at leaving college, but he attributed

that statement to my maidenly modesty, and in the end I walked off furious, if broken-hearted.[16]

In retrospect, however, Kate believed that she gained more than she lost by having to leave school. "Twice in my life," she said,

> I have gone down into the depths of despair because the way I wanted to go was suddenly shut off from me. Both times the way I had to go was the best possible way for me. Take this leaving of college: since I was nine I had been reading books on machines and engineering; my one year had given me the essentials of the profession. The rest I could do, and did do, for myself. My fierce determination to equal the young men I left at college served as a spur, and I worked with every bit of energy I possessed.[17]

Jim, too, suspended his high school education the next school year to join Kate and their father in the family business. William hoped to pay off his debts by liquidating his equipment, but there were no buyers willing to pay what the machines were worth. Selling at fire-sale prices would have left William debt-free, but the move would have wiped out all of his assets except patents. His only viable chance, he concluded, was to hang on until the market improved, when he would pay down his debt and work off his inventory before buying and building any new machines. William waited, the market improved, and the firm survived. By the spring of 1886, customer inquiries started to increase, and it seemed that the family business might start to prosper.

Kate spent long hours in the tool-building shop, but she had other interests, too, now that she was back in Rochester. On November 15, 1886, she was elected a member of the Fortnightly Ignorance Club, an organization that explored topics of interest to local professional and business women. Susan B. Anthony was a member and occasionally attended meetings with her sister Mary Stafford Anthony, a teacher and active advocate for women's rights. One of the club's founding members was Dr. Sarah Dolley, the third woman in the country to earn a medical degree and one of the intrepid women who had dared to register to vote with Anthony in 1872. Kate had much to share about her own experiences. At the January 31, 1887 meeting, she listened to a report of "Alaska and Its Resources" and "enlivened the Club with her piquant glimpse of a school girl's life at Cornell."[18] She continued to attend the club's evening meetings on and off that year.

She was also enjoying an active social life with her best friend, Emma

Michel, whose German-born father manufactured machinery in the Brown's Race area, near the Gleason Company. Emma's mother had died in 1883, and her father remarried in 1887. The Gleasons embraced Emma, an only child, as one of their own. It was a rare Sunday when she did not join the family for supper, and she and Kate's mother were especially close.

In January 1888, Kate recounted her adventures with Emma in a letter to her Rochester friend and Platt Street neighbor George Tegg, who was studying in England:

> I am now sufficiently recovered from my delirium of joy at receiving your letter to attempt an answer. There is such a power of news to be told I hardly know where to begin ... Emma and I went up to Ellicottville, 100 miles, to visit Jim McMahon, [a mutual friend] we staid three days and had a very high time; he and a friend of his took us down to Bradford to see the oil wells. Never saw such hospitable people as those oil country people are. We went to a ball that was decidedly different from any ball I ever attended before. I danced with a senator, a congressman, two drummers, the Chief of Police, and a farmer with his trousers inside his boots beside the young fellows of the town who were just like ordinary beings. The style of swinging is to take a good grab of a girl's waist and spin round; this was new to me and the first time it was tried on me I asked the man not to put his arm around me; five minutes after he was introduced as a cousin of mine. Almost everybody up there is some relation to us ...
>
> If you answer this epistle some time before next June, you can address your letter "Sage College" Ithaca, N.Y. for I am going back to Cornell next week for the rest of the year.[19]

In the spring of 1888, two and a half years after she dropped out of college, Kate returned to Cornell as a special student, no longer enrolled in the degree program she had entered in 1884. She fell ill soon after starting classes, but her father encouraged her to remain in school. "It is my wish," he told Kate, "that you stay the term out. I think that your health will be benefited by it. Besides it will be more satisfaction to you to close with the class with which you entered. Jim will commence to prepare as soon as you return. I will try and let him have two years. I am now going to make a statement that you may not like but when you compare the figures I think you will approve off it."[20] William explained that he had sold a right-of-way and water right, paid off some debt, and purchased a lot. He gave Kate all of the figures and his reasoning behind the transactions, adding, "I would feel very badly if I thought you worried over this."

By the middle of May, Kate was feeling so well she started playing sports. On May 12, 1888, she wrote Jim from the Sibley College drafting room:

> The fact that I play tennis so much worse that any one else down here is a sorrow to me and one of the girls suggested that the best way for me to learn was to begin some morning and play all day even if I had to be carried in on a stretcher at night; that to practice half an hour at a time wouldn't do me any good so I dedicated this day to the effort. Began with Mr. Dix at six o'clock this morning—he was exhausted by eight. Then I captured Bryant Blood and he played until after ten. Then Vivian Gay who had also been playing since early dawn for the same laudable purpose that I had, began to play with me. After four hours steady playing she was somewhat tired and as she is a wealthy young lady she hired a small boy to 'shag' her balls but I was as fresh as ever yet and did my own 'shagging.' We were just getting on in great style when a thunder shower came up and now its raining as though it had come for a week so I came over to do some drafting but have changed my mind to write to you ...
>
> I am more determined than ever to lead a life of unmitigated toil when I come back ...
>
> Lovingly yours,
Kate[21]

By May 20, 1888, Kate was looking forward to coming home, and her mind was on business. She wrote Jim:

> I hope we will get some more orders because we usually have to take in enough in May to last through the summer and those on the books when I left must be pretty well filled.
>
> What do you think of my coming home? You know I'm perfectly ready to come any time and if it would be apt to please the folks much more to see me now than a month from now—I would like to know it.
>
> I was just thinking this afternoon that it might have been better for me to take the course Mamma recommended—spend those three years at a fashionable young ladies school where I might have acquired a polish that would stand me in better stead when meeting customers than the solid knowledge I have gained here.[22]

A little homesickness lurks between the lines; Kate may have had difficulty picking up where she left off. She was ready to move on with her life. Kate left Cornell at the end of the academic year, returned to Rochester, and never completed her university degree.

NOTES

1   Eleanor Gleason, speech, January 29, 1931.

2   Obituary of William Gleason, *Rochester Post Express*, May 25, 1922, 22.

3   Bennett, "Kate Gleason's Adventures," 168.

4   Ibid., 169.

5   Although Helen Christine Bennett listed Kate's age at graduation as sixteen, Kate's name did not appear among the graduates in any year.

6   In the 1870s, German was taught in many public schools. Kate must have learned to speak it rather well, or wanted to master the language, because she belonged to a German Club in her twenties. She wrote to her brother Jim at some length about the club when he was away at Cornell University. At some time, Kate's language skills also included French, and both languages were very useful to her later in life.

7   *San Francisco Bulletin* (published as *Daily Evening Bulletin*), June 19, 1885 (vol. 60, iss. 63): 1.

8   "It all started with Kate Gleason, who made history for women engineers and was ASME's first woman member. She was admitted as a special student in mechanical arts at Cornell University where she studied from 1884 to 1885 and again in 1888. At Cornell she enrolled in the Sibley College of Engineering and was the first of what later became known as the "Sibley Sues." Janet Kotel, "The Ms. Factor in ASME," *Mechanical Engineering* 95, no. 7 (July 1973): 10.

9   James Gleason to Kate Gleason, September 24, 1884.

10   William Gleason to Kate Gleason, November 11, 1884.

11   Conable, *Women at Cornell*, 74, 78.

12   William Gleason to Kate Gleason, February 1, 1885.

13   Ibid., April 10, 1885.

14   Ibid., May 10, 1885.

15   Ibid., May 14, 1885.

16   Bennett, "Kate Gleason's Adventures," 169.

17   Ibid.

18   Fortnightly Ignorance Club minutes, Rare Books & Special Collections, Rush Rhees Library, University of Rochester.

19   Kate Gleason to George Tegg, mid-January, 1888.

20   William Gleason to Kate Gleason, March 30, 1888.

21   Kate Gleason to Jim Gleason, May 12, 1888.

22   Ibid., May 20, 1888.

CHAPTER 3

The Office Boss of the Gleason Iron Works

O N SEPTEMBER 18, 1888, all the shareholders of the Gleason firm—William Gleason, Amos Walder, Kate Gleason, Alexander Allan, Ferdinand Schwab, John Lauth, and James Henry—gathered to adopt bylaws for a newly formed enterprise, the Genesee Foundry Company, and all were named directors. Certificates indicate that there were four hundred shares at fifty dollars each, and the capital stock of twenty thousand dollars was fully paid. Two stock certificates were subsequently issued on October 9: number one to William Gleason, for one hundred shares, and number two to Kate Gleason, for twenty shares. The other five employee-directors were granted shares in 1889 or 1890.

In September, Jim traveled to Cornell for his entrance examinations, and Kate, who jokingly referred to herself as "the office boss of the Gleason Iron Works," began a steady correspondence with her brother. Three days after the shareholder meeting, she wrote:

> Dear Jim, Just about this time you must be wrestling with that arithmetic examination,—you want to pass that whatever you do and then it won't seem so bad if you are 'busted' on all the others. Papa says that he will not blame you if you can't get through the examinations because of course with the little time for preparation you've had, you have not had a fair chance.
>
> Don't forget the laundry I recommended to you. Here is a diagram, showing the way to find it,—the others are too steep in price and don't do things up so well. Remember to send me the question papers, if you mail em Sunday I can tell pretty well whether you are going in or not without waiting to hear from the examination committee.
>
> Lovingly, Kate[1]

September 24, 1888

My Dear Jim,

Your card was received Saturday and it lifted a two ton pressure off my mind. If you passed the three examinations of the first day you will probably get in any way. By the way wasn't Grammar one of them [?] You don't mean to insinuate, do you, that you have passed Grammar? If you have there won't be anything in this town too good for you.

The German Club was at our house Saturday night. I had forgotten that it was coming and when George Fleckenstein arrived I was sitting in the library with disheveled locks and a dirty face trying to make up my mind whether I would go to bed or go over to see Emma. Luckily he came some time ahead of the others so I had time to slick up. Fleckenstein is very agreeable, he speaks German well and he has considerable to say for himself in English too. He is quite exquisite, wore a ruffled shirt that came out of his vest in a regular puff. Vicinus doesn't curl his moustache at the ends any more and he doesn't plaster his hair low down on his forehead or part it in the middle. If it were not for his high hat and his russet leather shoes, I could have doubted whether it were him.[2]

October 2, 1888

Dear Jim,

... Papa is going down to Connecticut to-night and I will run the shop alone some more. I am getting so much practice at manageing this business by myself that pretty soon I will be able to let you and papa bend your mighty energies to running for aldermen or governors while I rake in the shekels ... I think of those girls I send you letters to, you will like Marian Colt and Nellie Lamson best. Marian Colt was my favorite but Nellie Lamson was a more general belle. You better call on them pretty soon for it must be pretty nearly time for the 'hop' and you musn't miss an invitation to that occasion to sport your dress suit.

As I didn't succeed in making out all the bills yesterday and it is one of my firm principles to send the bills out as near the first of the month as possible,—I will have to cut off the flow of my eloquence and go to work at them.

Lovingly yours,
Kate Gleason[3]

The Kidd Foundry & Steam Engine Works, Rochester, N.Y., 1865.

William Gleason at machine, circa 1888.

A youthful Ellen McDermot Gleason. Original portrait at the Susan B. Anthony House, circa 1870s.

James and Eleanor Gleason in front of William Gleason's home on Platt Street, circa 1885.

Gleason family portrait, mid-1880s. Left to right: James, Ellen, Eleanor, William, Andrew, Kate.

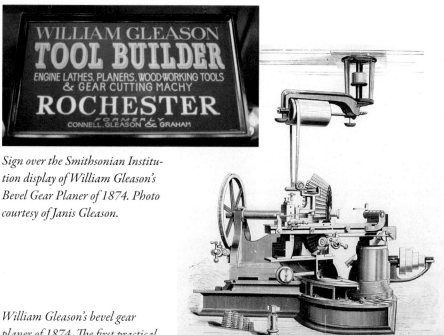

Sign over the Smithsonian Institution display of William Gleason's Bevel Gear Planer of 1874. Photo courtesy of Janis Gleason.

GEAR PLANER.
SHOWING WOOD-CUTTING ATTACHMENT.

William Gleason's bevel gear planer of 1874. The first practical machine designed for manufacturing the bevel gears.

Kate Gleason as a teenager in the early 1880s.

Kate Gleason in 1885.

Portrait of William Gleason, 1900.

Andrew and Edith Gleason, circa 1906.
Photo courtesy of Julia Gleason Rhoads.

Gleason employees in 1886. Kate Gleason is at the far right. James E. Gleason is in the upper row, fifth from left.

The main assembly floor of the William Gleason plant at Brown's race, 1891.

Kate at Clones, 1918.

Entry way into Kate's house.

Kate reading a letter, circa 1920s.

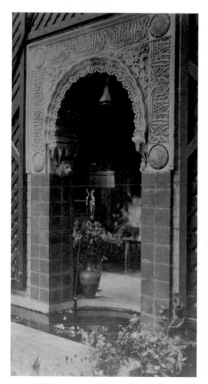

View from courtyard into living room.

The living room at Clones.

The courtyard at Clones.

Kate at Clones, 1915.

Hallway at Clones.

The courtyard looking toward the entrance and reception room. Gordon and Madden, Architects.

The front entrance of Clones. Gordon and Madden, Architects.

View of the entrance stairway and courtyard looking toward the dining room. Gordon and Madden, Architects.

View of Clones from the southwest. Gordon and Madden, Architects.

Dr. Dalley.   Presbyterian
Mrs. O'Connor   Congregationalist
Mrs. Parker   Episcopalian
Mrs. McKelvey   Unitarian
Mrs. Murphy   Roman Catholic.
Madge Vaswell   Agnostic.
Charlotte Smith   Episcopalian (Lutheran)
Beth Boynton   Episcopalian
Rosa Halyer   Unitarian
Mr. Ely   Congregationalist
Wm. Gladstone   Episcopalian
Dr. Rumford   Episcopalian
Prof. Thile   Agnostic
Dr. Schumann   Baptist.
Henry V. Sage   Congregationalist.
Edwin O. Sage   Baptist.
Dr. Doty.   Episcopalian
Miss Hall   Episcopalian
Patrick Mahon   Roman Catholic.
Mrs. Knickling   Baptist
7 Epis. 2 R.C. 3 Cong. 2 Unis. 3 Baptists 2 Han. 1 Pres.

Pages from Kate's black book. List of Kate's friends and their religious affiliation.

Names suggested for our
Progressive Euchre Club. Apr. 17, 1888.
Emma Michel, Frank Bryan, Louis
Beatty, George Tygg, Jim Gleason,
Kate Gleason, Walter Shawley, Minnie
Shawley, Ed Gordon, Edith Olmstead, Nick
West, John Nugent, Frank Nugent,
Geitie Nugent, Emma Joyce, Lillie Fawcet,
R. Marvin, Minnie Birkenhead, Ed Dempsey
J. Dempsey, Maggie Dempsey, Will Graham
Julia Barrett, Grace Reynolds, Maggie Whalen,
Katie Burns.

64" x 32" 40 % off 4 mos.
Stearns Manf Co. 60" Planer Mr. Barnhurst.
Hinman Engine Co. 36" Planer right away.

Scheme for dancing class, to hire
musicians estimate - $60
to hire hall estimate.   36
about $ a piece.   96

June 23rd. Cast iron resolution not to paint
my face

Total of inquiries in dollars. Monday May 10. 3500
                   Sunday May 9. 1765
                   Monday May 11 2400
                   Wednesday May 12 15-1-0
                   Saturday May 15 6350
                   Monday May 17 1650
                              800
                   Tuesday May 18 2500
                   Wednesday May 19 2790
                   Thursday May 20 910
                   Friday May 21. 2000

'86

3.00 Jul 7. C.C. JERSEY 3
2.50 4½ yds. cloth.
.10 8 2 yds. green velvet ribbon
.10 curled hair
1.50 9 1 pr. low shoes.
.03 1 rose bud. 2
.03 12 1 rose bud
7.26
.15 15 1 lb. raisins
.25 17 Windsor beach ticket 6
14 central ticket
.05 street car fare
.33 20 train & steamer to Windsor beach
.92 .40 22 one corset cover
.25 1 yd. ribbon, fringe for fan
.03 1 " cord. lacing " shoes.
.95 26 "Lalla Rookh" for Emma.
.05 street car fare
.05 27. C.H. mite box.
1.73

List of potential business inquiries followed by a list of personal purchases.

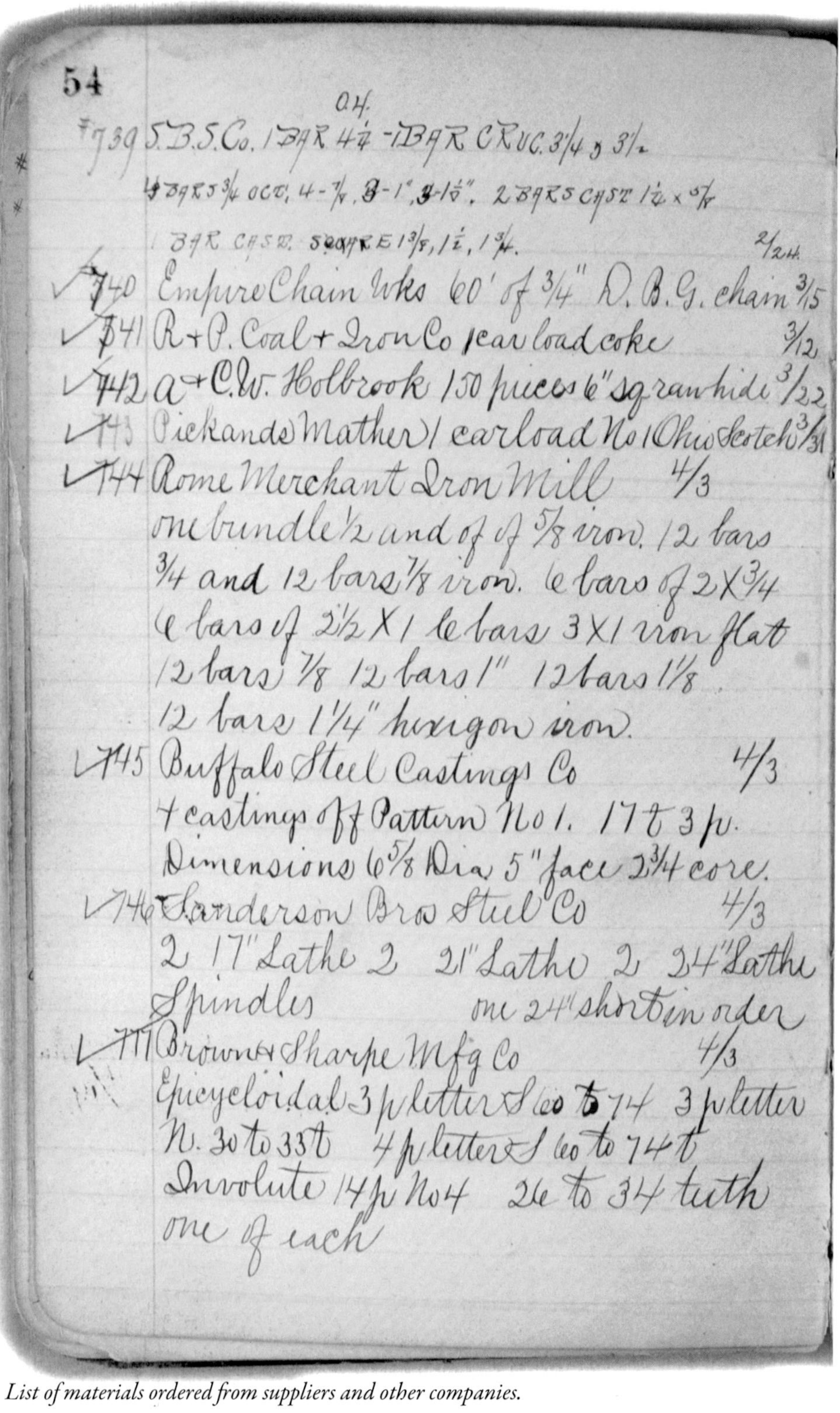

54

04.

#739 S.B.S. Co. 1 BAR 4 1/4 - 1 BAR CRUC. 3/4 x 3 1/2

4 BARS 3/4 OCT. 4 - 7/8, 3 - 1", 3 - 1 1/2". 2 BARS CAST 1 1/2 x 5/8

1 BAR CAST. SQUARE 1 3/8, 1 1/2, 1 3/4. 2/24.

✓740 Empire Chain Wks 60' of 3/4" D.B.G. chain 3/5

✓741 R+P. Coal + Iron Co 1 car load coke 3/12

✓742 A + C.W. Holbrook 150 pieces 6" sq rawhide 3/22

✓743 Pickands Mather 1 carload No 1 Ohio Scotch 3/31

✓744 Rome Merchant Iron Mill 4/3
one bundle 1/2 and of of 5/8 iron. 12 bars
3/4 and 12 bars 7/8 iron. 6 bars of 2 X 3/4
6 bars of 2 1/2 X 1 6 bars 3 X 1 iron flat
12 bars 7/8 12 bars 1" 12 bars 1 1/8
12 bars 1 1/4" hexigon iron.

✓745 Buffalo Steel Castings Co 4/3
4 castings off Pattern No 1. 17 t 3 p.
Dimensions 6 5/8 Dia 5" face 2 3/4 core.

✓746 Sanderson Bros Steel Co 4/3
2 17" Lathe 2 21" Lathe 2 24" Lathe
Spindles one 24" short in order

✓747 Brown & Sharpe Mfg Co 4/3
Epicycloidal 3 p letter S 60 to 74 3 p letter
N. 30 to 33 t 4 p letter S 60 to 74 t
Involute 14 p No 4 26 to 34 teeth
one of each

List of materials ordered from suppliers and other companies.

778 Buffalo Steel castings Co   4/5
4 castings off Pattern No 2   18℔ 1¼ p
8¼ X 7⅜ face 3" core

779 Sanderson Bro' Steel Co
3 24" Lathe Spindles   4/9

780 P. Dutcher + Co
one carload molding sand 4/11

781 Collord + McKeefrey Florence iron.
One car April Delivery
    "   "   May   "
One note Apr 30   Other May 31,

782 Kearney + Foot #/o Chambers St Patterson N.J.
4 dozen 14" hand smooth one safe edge

783 Hussey Howe + Co New York
6 bars of 2" + 2½" and 3 bars of 3"
Mchy steel. 6 bars each of 1⅜" + 2½"
and 8 bars of 1¼" Steel. S of 8,

784 E. D. Ransom + Co
one barrel of ceylon lead

785 Westcott chuck Co   for Davin + McClure
One 15" X 4 jaw chuck independent 4/19 4/24

786 One car load Hubbard Scotch.
Andrews + Hitchcock

	GERMANY & AUSTRIA.	REST OF EUROPE	TOTAL.
1893	1150.00		
94	1150.00	1200.00	2350.00
95	1075.00		
96		2650.00	
97	2975.00	10550.00	13525.00
98	10135.00	18250.00	28385.00
99	14100.00	9600.00	23700.00
1900	13600.00	32300.00	45900.00
01	8700.00	18600.00	27300.00
02	1000.00	16000.00	17000.00
03	7800.00	27300.00	35100.00
04	1100.00	27000.00	28100.00
05	8900.00	33300.00	42200.00
06	11300.00	22800.00	34100.00
07	24700.00	16400.00	41100.00
08	17700.00	18200.00	35900.00
09	3300.00	5600.00	8900.00
1910	4900.00	9000.00	13900.00
1911	11800.00	49,700.00	61500.00

List of product values sold to European countries from 1893 to 1911.

Capital	Profit	
114,000.00	4400	'95
117,600.00	8000	96
120,500.00	4900	97
123,700.00	10400	98
135,800.00	17300	99
137,900.00	20800	'00
153,200.00	19700	'01
185,000.00	30000	'02
195,200.00	36000	'03
201,000.00	26000	'04
213,700.00	43500	'05
273,300.00	108300	06
465,700.00	136600	07
472,000.00	44300	08
455,000.00	126800	Jan 1910
600,000.00	160,000	'11
580,000.00	86,000	12
645,600.00	113,000	13

Capital and profit figures by year from 1895 to 1913.

CONTRIBUTION WOMAN SUFFRAGE

Feb 1904	W. A.+ 25 oo N. A. 25	50.00
Oct. 11 1905	" 25 " 25	50.00
Feb 2 1906	ANTHONY BIRTHDAY RECEPTION	75.00
Mar. 16 1907	K. A. OREGON CAMPAIGN	25.00
Nov. 30	W. A.+ 25 N. A.+ 25	50.00
1908	K. G.	100.00
May 4 1909	K. G.+ 50 W. G.+ 25 oo	75.00
Apr. 2 1910	" 50 " 25 oo	75.00
Oct. 15	Mrs. Bishop's Reading	140.00
" 1911	K. G. California Campaign	125.00
Feb. 10 1912	K. G. N. J. Organizers	25.00
	W. G. to Anna Shaw	50.00
Mar. 7	K. G. to Sylvia Pankhurst.	15.00
Apr. 16	Campaign 6 states (engagement)	100.00
June 3	" 6 "	100.00
6	" Ohio (Mother's birthday)	267.99
Jul. 4	" "	500.00
Sept. 4	Roch. Local + 10 oo Ohio + 10	.20 00
Dec. 1	100 per month National, 25 per month Local	1500.00

Contributions made by Kate and William for woman suffrage.

76) 69.22 (4.26

.15	OCT. 25	ONE TOOTH BRUSH	81.37
			2.00
.25	30	SULPHATE OF CALCIUM	1.50
.40			1.00
.05	31	CHURCH COLLECTION	1.20
			89.07
.65	NOV. 5	REPAIRS ON SHOES	.75
.70			2.00
.22	8	LINEN FOR MAMMA'S HANDKERCHIEF	91.82
5.00	9	PAIR OF FINE KID SHOES	
.60		INDIAN CLUBS	
.25	10	SULPHATE CALCIUM	
6.07			
1.25	15	DYEING 'GYM' SUIT	
.12	17	BRAID FOR 'GYM' SUIT	
1.37			
1.20	22	"LITTLE WOMEN" FOR ANDREW	
.25		SULPHATE CALCIUM	
.30	26	RIBBON & SILK, MAMA'S SCARF	
1.75			
.25	DEC. 2	SULPHATE CALCIUM	
1.00	9	R.R. TICKETS, EMMA & I	
1.25	11	TENNIS SHOES	
.09	"	1 'αρδ Hoρδ	
.11	"	δαres Hαι ραιsιrs	
.10	"	σHρIM	
2.47			
.10	16	δερ παρασιτε	
.30		βοξ παπερ	
1.95	17	πραιρ βοοκ παρ έιι	
2.35			
2.50	23	σοκκιιν βαν. μrs ΗιΜβερλι	
.50		διο'es καρ Ἐλιτορ	
3.00			

List of personal items purchased by Kate.

Total Wages 1/20 1900 - 4/1901 1/20 43,524.96

Shop Work do. 12×248.79

31,276.17

Expense percentage April. 20. 1901: Labor 12,248.79

Expense 2,700.00

Int. 3 Cty. 2,000.00

Ind. 3 C. 1,600.00

Manf. 5,000.00

31,276.17) 23,543.79 (75 %

Oct. 20, 1904

			WT.	WAGES.
FOUNDRY CASTINGS JULI - DEC I, 1905 DAILY AVER.			10040#	122.75
" " " " 1906 " "			16800#	228.95

SINKING FUND FOR ACCIDENTS. AMTS. Jan 1, 1907 = #13000

Average wages common Labor in mach shop July 3 - 11   .22

Miscellaneous business notes.

Average wages Machinists & apprentices Apr. 7-1900 .219

  94 machinists 14 apprentices.

Average wages 92 machinists 13 apprentices Apr. 20,1901 .235

„	„	78 journeymen 16 „	„ 20,1902	.239
„	„		„ 20 1905	.256
„	„	all not on salary	Mar 23 1907	.254
„	„	„ „ „	Sept 24-07	.261
„	„	„ „ „ „	May 14-09	.263
„	„	„ „ „ „ „ „	Sept 4-09	.272
„	„	338 men	Dec 15-09	.265
„	„	390	April 2-10	.266
„	„	349	March 1-11	.281
„	„	330	July 3-11	.29
		303	Dec 30-11	.295

Number of employees and their average wages.

Sept 28 '99 superior right, bicusped anterior approximate Gold = 3.50

Feb 6 1900 ,, superior 12 year molar, coronal, distal angle Amalgam 2.00

,, 8 ,, ,, 1st bicusped distal   Gold 4.00

Jan 27 left ,, 1st ,, gold crown superseded by porcelain faced crown 12.00

,, 28 ,, ,, 2nd molar occluding surface Amalgam 1.00

1901 Jan 21 ,, lower 2nd bicusped distal surface silver 1.00

Mar 19 1902 ,, superior 2nd bicusped pivot tooth superseded by Logan crown 10.00

May 7 1903 ,, lower 2nd Bicusped {2 cement renewals 2.50 {1 gutta percha new cavity

,, 29 right ,, ,, ,, distal approx. coronal cement 1.00

Aug. 7 left ,, ,, MOLAR   GUTTA PERCHA 1.00

,, ,, upper ,, ,, ,, 1.00

1904 Mar 7 right ,, ,, BICUSPED MESIOCORONAL APPROX CEMENT 1.00

,, 28 left ,, ,, CUSPID MEDIAL ,, 1.00

,, ,, ,, ,, LATERAL DISTAL ,, 1.00

JUL 26 ,, LOWER 2ND BICUSPED DISTAL SURFACE NERVE KILLED, CEMENT (BUNBURY) 1.00

Aug. RIGHT UPPER   ENAMEL ,, 3.00

Dec 2 LEFT LOWER 2ND MOLAR (REFILLED) GUTTA PERCHA 1.00

1905 RIGHT ,, ,, BICUSPED ,, CEMENT 1.00

FEB. 14 LEFT UPPER CUSPID MEDIAL ENAMEL 5.00

,, ,, LATERAL DISTAL ,, 5.00

MAY UPPER R. CANINE CEMENT BRININGTON 1.00

,, R. 2ND BICUSP. ,, 1.00

LOWER R. 2ND ,, ,, 1.00

,, R. 2ND MOLAR GOLD 2.50

,, L 2ND BICUSP LOGAN 8.00

List of Kate's dental work.

Low water mark. June 22, 1899.

Shop Mortgage	16,900
Bank	6,000
Whittlesey	4,000
Irwin note	500
Fdy. notes	7056.58
Current	1450.00
Castings	1069.33
Irwin	3367.96
House mortgage	2500.00
	73,147.87

Low water mark Jul 15, 1905

Brown's Race mortgage	36,000
share University Av. "	17,290
Bank	8,000
Fdy note	10,000
Current	2,000
	73,290

A comparison of business years between 1899 and 1905.

"OUR LIST"

"A" AIKENHEAD, M.; ALLEN, M. ANTHONY. B.
AIKENHEAD, R. ASHLEY. C.
ALT, F.
ALEXANDER, C.

"B" BRYAN, F.; BLACKALL, G.; BLACKALL, R. BYGRAVES. N.
BEATTIE, L.; BOSTWICK, B.; BOSTWICK, C. BOLGER. W
BEATTIE, F.; BRICKNER, I.; BOYD, E. BRYAN, H.
BRIGGS, I.; BROWN, E.; BEAISAN. C. BIRDSALL, E.
BOWERMAN, I.; BRITTON. A.; C. BARTON.; BUSBY. M.
BARTHOLEMEN, M.

"C" CUNNINGHAM, F. CAMPBELL, A. CLEMENTS. M.
CLEMENTS, I. CONNOLLY. D. CONNOLLY. F. CONNOLLY. I.
CRONK. W.; CHARTRAIN, L.; CURRIER. N.; CHAMBERLAIN L
CORNELL, L.; COLE. C.; CHITTENDEN. L.A.; CROSBY. M.

"D" DENGLER. W.; DANFORD. E. DANA. C. DARNELL, H.
DEMPSEY, M. DEAN, F.; DEVERAUX. H.; DURNEY. E.
DEMPSEY, E.; DRAKE. C.
DEMPSEY, I.

"E" ELWOOD. C.; ELBS. I.
ELWOOD. F.
ELWELL, G.

"F" FAWCETT, L. FRENCH. G.
FOX. A.
FOX. E.
FENNER. B.

First page of alphabetical list of Kate's friends.

October 4, 1888

Dear Jim,

... As a natural consequence of the weather for the last couple of weeks, the foundry yard is a sea of mud and the men pole themselves across on boards in order to navigate it. Dan Burns is getting so far along that he brings his bottle of whiskey to work with him and then he and Shippy enjoy themselves with it when people aren't watching them. The finances of this concern won't stand hiring a detective to look after them and as the manager of the shebang just now, I have instructed Marriott to discharge Burns the next time he is caught and told the Foundry Co. it would be well to demand Trihey's resignation but the foundry is used to such playful little streaks in its men and I suppose Mr. Allen thinks he would only be jumping from the frying pan into the fire by discharging him. If you come across any receipt [recipe] in your travels for making moulders keep sober and pay their debts I wish you would carefully preserve it. Andrew's getting on like a house on fire.

> Lovingly,
> Kate Gleason[4]

October 9, 1888

My Dear Boy,

I had a gorgeous time at the German Club Saturday night. In the first place there was a lot more gentlemen than ladies and then Henry Bartholomay was back and you know yourself how much more interesting things are when the object of your affections is around and then there was a Herr Dr. Satterlee who seemed to scorn me and my ways considerably so I whiled away considerable of the evening making life a burden to him until every time I would make a remark to him, a hunted look would creep into his eyes that was positively pitiable.[5]

With the hope that you will do nothing to tarnish the glorious name I left at Cornell but on the contrary will do everything in your power to add several shades of gilt to it, I remain

> Lovingly yours,
> Kate Gleason[6]

October 13, 1888

My Dear Jim,

I will transmit my thoughts to you through the medium of a lead-pencil because I am too comfortable to take my feet down from the chimney-piece as I should have to do if I used the caligraph ... We stand a show of receiv-

ing an order for a ten ft. square planer as well as some other tools from a
Cleveland concern ...

> With the usual amt. of affection,
> Kate Gleason[7]

Two days later, at the Fortnightly Ignorance Club, Kate described her
experiences wearing the reform dress of Mrs. Annie Jenness Miller of New
York. A physical culture author and publisher, Miller had created a style
of undergarments that simplified and reduced the heavy, constricting, and
distorting fashions of the day. Miller's system of "art in dress" featured fine
needlework, ornamentation, and looser styles that aimed to unburden the
female form to some degree. Kate declared that the garments were "comfortable and perfectly adapted to a business woman's needs." Her testi-
mony must have made an impression, because at a subsequent meeting in
January, the minutes reported, "Upon motion, [it was] resolved that Miss
Gleason be a committee of one to correspond with Mrs. Jenness Miller
with an eye to the emancipation of the Club from the problem of its present costume. Carried."[8]

On October 16, Kate again wrote to Jim, encouraging his relationship
with one of her favorite professors, Albert S. Smith, an assistant professor
of mechanical engineering. Smith had been earning his master's degree at
Cornell while Kate was there and later became a beloved dean of the Sibley
College of Engineering and acting president of the university.

> Your epistles to the head of the family and to the junior "boss" have been
> perused with a great deal of interest. It is a profound pleasure to hear that
> you have succeeded in "working" Prof. Smith so successfully. Only go on,
> dear boy, and you will yet be the makings of a machine tool agent.
>
> This morning Father (do you notice the high-toned appellations that I
> now give our parental relatives?) went to Wellsville. The McEwens wanted
> him to appraise some tools, I suppose they are going to desolve partnership,
> at least they didn't intimate that they wanted to trade for a shop ful of new
> ones. There is a man in the oil country ordered two taper lathes and a 30"
> planer. He traded in a house and lot in the city for them. So that now we are
> the proud possessors of a mansion on Glenwood Av. as well as our palatial
> residence on Platt Av.
>
> Connell and Dengler have taken their new 48" planer home and given
> us a note consequently our bank account this week is very prosperous. If
> you are intending to need much money in a hurry now's your time, maybe

we won't have any left next week. Last week we paid up for our stock in the foundry and when I walked into the German-American Bank to draw the cash on my personal check for $6,000 I felt as though I owned about one half to two thirds of the town. But I managed to conceal my emotion and passed in the check with a careless 'give me three of that, please' air and the cashier was so overcome he hadn't the courage to say anything but that he understood the climate in Canada is delightful ...

I find the German Club a very entertaining gathering these days. We have two new members, Miss Goetzman and Miss Bausch. I believe now that we have the cream of the German aristocracy in town. At least we have representatives of the biggest brewery and the biggest whiskey manufactory in Henry Bartholomay and Miss Goetzman. Alice Kaiser continues to distinguish herself. The Club had unanimously elected a very pretty Jewish girl, Claire Wolff. At the first meeting after her election, she couldn't come and Miss Kaiser took it upon herself to write her that therefore we had no more use for her. If Miss Kaiser only knew a little more than she does now she would know a little. Lillie's social club was inaugurated with a blaze of glory last night at Jennie's. They implored me with tears in their eyes to be temporary chairman or something so I proceeded to run things with my customary grace and ease. We elected a set of officers. Despite several hints that I threw out that the position of Treasurer would be highly acceptable they didn't elect me. How is it I wonder that I was never yet elected Treasurer of a club and I run for that office every time.

Yesterday afternoon I went up to see about joining the gymnasium in Wisner's Block and I have fallen a hopeless victim to the charms of Miss Carter, one of the proprietors thereof. She even induced me to promise that I would appear, at the first lesson next Friday afternoon, in a blouse and Turkish trousers. Gracious it's worse than a bathing suit. I have been put in the advanced class but it's my private opinion that I won't be there long for Miss Carter spoke of the girl's hanging by their toes from the rings as a very ordinary feat and as I can't be trusted to hang on to them with my hands even when I get "boosted" up to them with the aid of a step ladder, I am afraid the class will not consider me a shining light.[9]

On October 25, Kate recounted to Jim her pioneering experiences as a female traveling salesperson, out on the road selling machine tools.

Dear Jim,
I have just returned from a four days' drumming trip down in Ohio and in

the course of my travels I took an order for a large planer and saw the Cincin-nati Centennial Exposition. You see Father has been out of town so much lately that I thought it would be well for him to stay at home and get the run of the business while I went after this order. I communicated this idea of my own to Emma and on the strength of a chicken supper Sunday evening and our persuasive eloquence after it, he let me start early Monday morning.

I met Mr. Arthur Curtis on the train. He had taken an early start for the Cleveland convention of the Delta U's so that he could stop off for a day with a friend in Buffalo. I asked him not to tell you about my little trip for fear I wouldn't get the order in which case I would rather you remained in dense ignorance. You see I had a good many quakings of my valiant spirit. The customer's name was Rudolph Schneble. That sounded as if he might be a Dutchman and if he were an old Dutchman the chances were he would not like to have a fascinating young woman like me after his order for ma-chine tools and might put me out at the end of a shot gun. But fortunately for me Mr. Schneble is young, only 23, he has until lately attended a Jesuit College, is not used to girls so I managed to make a 'crush' on him quite early in the morning. After I secured his order I traveled around Dayton to call on all the firms I had ever heard of there and in every case I was treated with "the most distinguished consideration."[10]

Friday morning

The exposition was worth going to see. I admired a power of mchy [machin-ery], pictures, implements of war, defunct Indians, tested all the different samples of beer (you know Cincinnati is celebrated for its beer) and floated on the Canal in a Venetian gondola.

I called on a lot of people in Cincinnati, Lodge, Davis & Co. used me brown, took an hour and a half to show me everything in their works. I wouldn't take their business for a gift. They cater to a mighty cheap trade and even in the styles they do their work, I can't see where there's any mon-ey in it. This morning there was a man from Brown & Sharpe's in here who hired Prof. Albert Smith when he first left college. He said the usual thing of Prof. Smith,—that he is the most perfect gentleman he has ever met. Wish you could cultivate Prof. Smith, Jim, and see whether there is any get-ting at the secret of his charm. If you can find out whether he belongs to any denomination of Christianity, let me know.

Lovingly,
Kate[11]

It appears that the office boss of the Gleason Iron Works had honed the skill of communicating efficiently at the expense of being grammatical. Many of her letters to Jim run sentences and paragraphs together without benefit of punctuation. The letters do impart the sense that Kate was a busy woman who nonetheless made a point of keeping her brother informed of the news from Rochester. The fact that Jim carefully kept all of her letters to him in a file in his office leads one to believe that he appreciated receiving them. Unfortunately, any letters Kate may have written to her youngest brother, Andrew, during the years he was away at Cornell, have never surfaced.

NOTES

1   Kate Gleason to James Gleason, September 21, 1888.
2   Ibid., Sepember 24, 1888.
3   Ibid., October 2, 1888.
4   Ibid., October 4, 1888.
5   Henry Bartholomay was the head of the Bartholomay Brewing Company.
6   Kate Gleason to James Gleason, October 9, 1888.
7   Ibid., October 13, 1888.
8   Fortnightly Ignorance Club minutes, Rare Books & Special Collections, Rush Rhees Library, University of Rochester.
9   Kate Gleason to James Gleason, October 16, 1888.
10   Ibid., October 25, 1888.
11   Ibid., October 26, 1888.

CHAPTER 4

News from Home

LETTERS CONTINUED BETWEEN ROCHESTER AND ITHACA over the following weeks. William wrote Jim about the business, telling him that he looked forward to having his assistance. "I have been so long in the harness and always on the upgrade," William wrote, "that I will be more than willing to delegate a portion and eventually the whole off the load to you and Kate and Andrew. Business is good ..."[1]

Jim wrote Kate about a young man named Warner who was supposed to be his roommate before he got suspended for a year for "calling on one of the young ladies at Sage Cider Raid Night." Warner, it seems, had climbed a ladder to the dorm's second story and sat on a windowsill outside the coed's room, chatting with her and having a pleasant time.[2] In a subsequent letter, Jim added that Warner was moving to Rochester and would probably call on Kate because he knew about her glorious record at Cornell. "Now don't take him for a specimen of the fellows I go with for he ain't," Jim warned. "I picked him up for a roommate late in the term when I couldn't get anyone else. He is a good-hearted fellow but as poor a student as I ever saw and not over stylish. But say take him on a cane rush and he is a whooper."[3]

Kate, in turn, advised Jim against joining a fraternity, based on her own disappointing experience in a sorority. As freshmen, she and two other women had pledged the Psi Chapter of Kappa Kappa Gamma, established at Cornell the previous year, but as time went by, Kate found that she did not get on as well as she had hoped with her nine sorority sisters. She warned her brother against making a similar mistake.

27

November 8, 1888

Dear Jim,

... I hardly know what to think of the card you enclose. If it means that you have joined the Delta U's I have nothing to say. If it means you are thinking about it, I would advise you to wait until after Christmas as you said you would before you left home. You can't know, Jim, how the opinions about people you form in your first term will change and if there is one thing more than another that is a weariness of the flesh it is to belong to a college fraternity when the members are no longer congenial to you. However it's safer to join Delta U. in a hurry than Beta Theta Pi or even Si U. The Beta Thet's have too lately come from under the shadow of their past reputation to be beyond risk. And the Si U's comprise one or two of the meanest prigs in the University. If Philip Price Barton is still there it might interest you to notice him as an instance of how really disagreeable a person can become when he gives his whole mind to it. Once I had him for a partner at a progressive euchre party and he was so cold he made my teeth ache.

Lovingly,
Kate

[P.S.] Hope you will cultivate Prof. Smith all he'll let you. If you manage to acquire any of his charm of manner, I'll give you my blessing.[4]

The next night, Kate witnessed a tragedy. A raging fire destroyed the Steam Gauge and Lantern Works, located in Brown's Race near the Gleason shop, killing thirty-eight men who were trapped when the blaze engulfed the building's exit platform. Fires were common, but Kate experienced this one at close, personal range, and it deeply affected her.

November 10, 1888

Dear Jim,

I mail you to-day a paper with an account of the big fire we had down here last night. It was something terrible, Jim. The wind blew a gale and sent great sparks and chunks of burning wood all over the city. The sparks flew as far as Irwin's house, the blaze was so bright, you could see to read a newspaper by it at Mr. Walder's on Fulton Av. We had to have men posted all over our roofs at the foundry and threw great burning brands over the brink. If it were not for the rain that had soaked everything through, we would have been goners. The worst was the awful loss of life. We were late taking a heat off in the foundry and so all our moulders were over trying to catch the men who jumped from the building. To hear the poor creatures shriek and pray for help was heart-sickening and the awfully desperate chances they took in jumping made one almost frantic. Three men jumped from the third story at the back way down to the flats,—think of that. Two of them were killed by the fall but the other struck in a bed of mud and walked off. One poor little boy was high up in the building ten feet above the extension ladder. When he saw it couldn't reach him, he jumped, missed the life net and was dashed to death on the stones. I never want to see such awful things again. It makes a body's hair turn gray. When I came down to work this morning, I went to the back of the foundry to have a view of the ruins and down on the flats I saw poor old Mr. Stone who seems to have gone crazy looking for his son. You remember his daughter-in-law was murdered by a tramp about a year ago and now her husband is burned to death. The poor father was poking about on the flats, trying to wade through the icy stream that goes from our water wheel so as to get over to the Gorsline Flats and nothing anyone would say seemed to make any impression on him.

Lovingly yours,
Kate[5]

By November 14, however, her mood had lightened.

Your friend Warner has turned up—why didn't you warn me he had already reached town and would be liable to call at any moment? He came Saturday night when I wasn't expecting any callers,—my hair was disheveled as usual, there were several shades of coal dust and machine grease on my face and I had taken off my shoes to dry my little wet feet. Otherwise I was looking fine. With his usual thoughtfulness Andrew showed Mr. Warner right in the front parlor so that I didn't have any chance to escape through the window. However I don't feel as badly as I might for I see Mr. Warner has a great big splotch on his light derby and he can't be very particular or he wouldn't wear it. After I had embraced him tenderly and told him I would be a sister to him I retired to wash my face and put on a tea gown. We got on like a house on fire.—He told me what a great boy he is and how hard you study and various other interesting things.

Business seems to be booming as per agreement, inquiries are thick and orders are numerous so that on the whole I am glad I supported Harrison.

Lovingly,
Kate Gleason[6]

That same day Warner wrote a letter to Jim:

Dear Old Man,
Arrived about noon last Thursday ... I called on your sister last Saturday and the folks were tickled immensely at seeing 'Jim's chum.' I didn't tell them that you and I were only intending to room together.

Miss [Emma] Michels is a dandy. She leaves today for Detroit I think. I attended a euchre party at your house last night. Got taken suddenly rich and between that and trying to stay and make myself agreeable, I made a— fool of myself and suppose everybody is disgusted with me.

It's a deuce of a job to kill time. I don't have more than 12 hours a day though; I sleep the other twelve. Please excuse the looks of this sheet as I am hardly awake yet, though it is 2:15 in the afternoon. Write soon.

Chas. Warner[7]

On November 21, Kate wrote:

Dear Jim,
I'm still in the preliminary stages of gloomy despair for Emma may go to Detroit any day; but as she hadn't got there yet, I have not plunged into the dark abyss itself. You see her hostess, to be, is sick and until she recovers, Emma continues to illumine us with her society. But Mrs. Schlick is now convalescing and may be expected to send for her any day.

Mother is going down [to Cornell] next week sure, Elinor[8] is going too, of course, and she has a couple of stunning new gowns and a jacket. Mother hasn't any new gown and what's worse, she has resurrected the old tweed dress I wore to the office for the last three years, washed it and had it made over for herself. But I pledge you my honor, Jim, that I will use every effort to secrete the thing so that she cannot take it with her as she now intends doing. If I manage to impound that, she will have to wear her best dresses and then she can't miss making a captive of every boy down there.

Warner is what you used to eloquently term 'a lallah,' isn't he? His Father wrote him that he would have to support himself for a year and if he managed to do it and would promise to be a good boy, he might send him through college another year. So he came down to ask my advice about looking up a job. He can't afford to take the job of apprentice here for he couldn't support himself on the wages, $4 a week. His Father sent him a letter of introduction to D. A. Woodbury but when young Warner presented it, Mr. Woodbury told him (so young Warner confidingly imparted to me) that he considered Mr. Warner senior had swindled him about a $30,000 piece of property and he would like to see him and all his relatives including our young friend in,—well in a warmer climate, before he would help him one cents' worth. Warner was down here Friday afternoon, and Saturday afternoon and Monday afternoon and then again Tuesday afternoon,—he hasn't showed up yet to-day but perhaps I can still support life, even if he doesn't come. There is a good deal of sameness in his conversation,—it is principally monosyllabic unless he is talking on the merits of '92 in cane rushes, base-ball etc. etc. as compared with '91 ...

Business is good, in fact it is very good and Father has about made up his mind that you are to stay long enough to graduate; you ought to make a violent effort to do it in three years for we are only waiting for your return to spread out like the American Eagle until the shadow of the Gleason tools extends around the earth. We are making a gear planer for the French Government now. I'm glad you have asked Madge to the party and hope you'll enjoy yourself.

K.G.[9]

Rochester, N.Y., Nov. 26, 1888

Dear Jim,

... Warner is coming in the shop to learn the trade at the munificent salary of $4 a week. Neither Father or I like him but we thought we would give him a chance, however I don't expect he will improve it for he doesn't seem to know enough to come in out of the wet.

There is no use trying to be so fearfully economical as to get a room for a dollar a week. You better introduce young Ide to Mother and if she likes him, it would be best I think to go in with him on the $4 room. His Father's reputation in business is good, I've heard of him more or less since I've been down here and they say children take after their parents so the chances for Ide being about right is better than Warner's,—if Warner's governor is what we hear.

Emma and I have both resigned from the Dawn Club. She resigned because she expected to go to Detroit and I resigned because I don't care to go out any more than is necessary this winter and only joined the club because she wanted me to. I haven't heard whether it busted up after we left or not. It very likely did, for I don't see how any club could get [on] without us.

Yours lovingly,
Kate[10]

Rochester, NY. Dec. 7, 1888

My Dear Jim,

Emma ran the house in fine shape during the absence of the head of the family. You must never divulge the awful secret but the fact is we had corned beef and cabbage and we still live. If Mother only knew we had had some of that dyspeptic stuff she probably wouldn't sleep for a week.

Andrew has been promoted to the handle lathe and Warner has been promoted from scraping lathe frames for Palmer to the gear cutters vice Andrew resigned. Palmer told Dick that if Warner wasn't put in some part of the shop where he couldn't see him, he would murder the poor boy. Since Warner went on the gear cutters Monday he has spoiled two gears and Andrew only spoiled one all the time he was on. Warner's Father has sent him some cash and in consequence he has laid off since yesterday morning and I believe he now intends to buy something beside that spring over-coat and he may even invest in a cake of soap of his own. I will save up all the good advice I have for you until I see you in two weeks. I have several new and fresh brands that I have never given you yet.

Lovingly,

Kate[11]

In an undated letter, Jim told Kate about her stellar reputation at Cornell:

One thing I forgot to mention, it is a conversation I had with a Miss Porter recently; it shows very well the interest with which some people watch your career. She was very anxious for a full account of your business relations, and I proceeded to make her eyes bung out with facts and never appreciated before myself what a gem you are of late. She asked if you ever made any individual business ventures. I said Miss Kate Gleason is a stockholder in the Genesee Foundry Co. and Secretary of that concern. In fact gave a full account of you including your being a member of the Chamber of Commerce which I take it from one of your letters you are. Perhaps I should not have taken it as I did though. Miss Porter objected to your having the handle Miss which I assured her did not appear on the bill heads. From her talk she intends to burst soon on the business world as Porter & Father of Buffalo. I predict that she will never be a rival of yours in fame.

JEG[12]

Jim came home for the holidays, and his correspondence with Kate continued after the New Year.

Rochester, N.Y. January 6, 1889

Dear Jim,

... How does trigonometry strike you? Let me know what sort of lick you have on the prelims.

I am plunged in considerable gloom myself just now and when I tell you that in one short twenty-four hours I lost you and Emma and Grover you can form some idea of the depth of the gloom above mentioned. Elinor took Grover [Kate's dog] out for a walk Thursday against my express orders for I had become tired of having my mind all harrowed up with the times she managed to lose him. But she took him out just the same and when she returned she left him outside to gambol about the yard and he protracted the gamboling until when I came home to tea, he had been gone a couple of hours. I was pretty tired on account of a whist-party at Miss Hall's the night before together with my grief at loseing you and Emma so I went to bed directly without waiting up until he should come home. And he didn't come home so the next day at noon I went over to see if he had returned to University Av. I found he had come there the day before and Jacobi probably thinking I was tired of him had taken him out of town with him without going through the little formality of leaving the $15 I had paid for him behind. The landlady thought it was pretty mean and said she had offered to keep the dog a couple of weeks if necessary until I should call for him but Jacobi wouldn't leave him. So I went down to call on Judge Keeler. You know he hasn't made a "crush" on me yet and I guess I must be the only girl in the cathedral parish who has not been smitten with his charms so I smiled on him enough to make him think I was just about to be added to his list of victims and he promised to arrest Jacobi and [get] the dog for me if they could possibly be found. But I havn't much hope.

Lovingly,
Kate[13]

Rochester, N.Y., Jan'y 9, 1889

Dear Jim,

I've heard from Emma and she is prospering. She has taken first prize at a euchre party and beside the glory it is a very valuable prize. And at a dancing party she had enough partners and to spare. She writes me that she can't stay there long, her clothes won't do and she has sent on orders to her dressmaker here to make up her wedding dress (the one Mrs. Michel gave her [for] Christmas) and to make it décolleté, the décolletér the better.

Andrew finished up six worms [gears] yesterday, a job that heretofore has been entrusted to no one but Doyle. But Doyle was busy so Father showed Andrew how to do one and then he finished up the other five all alone by himself and they are a credit to the establishment.

Very lovingly,
Kate[14]

Rochester, N.Y. Jan'y 30, 1889

Dear Jim,

Just now I am doing a land-office business between the machine shop work and some arrangements I am trying to make to have Mrs. Jenness Miller come to town to lecture before our club. I am enjoying life exceedingly just now, strange as it may seem with you and Emma both gone and I haven't found a suitable young lady to cultivate as I intended in Emma's absence either but I have made deep impressions on two young men or else they have succeeded in making me believe I have and you know yourself that a thing of that kind adds considerable interest to life. I can't tell you about these "crushes" in detail as it wouldn't sound well in print even if you should find it interesting which I doubt. We had a party at our house last night, Mother's Euchre Club, and in preparation she had taken up most of the carpets, had the walls re-papered, the curtains laundered and had to put back most of the things she tore up so that the house looked very nearly as well as before she began to make the preparations.[15]

Rochester, N.Y., Febr'y 5th, 1889

Dear Jim,

Business doesn't show any sign of slacking up yet and I hope it won't and then next fall we will have a boom and get rich. We are building some tools for Sargeant and Greenleaf and the way Mr. Sargeant comes down here and keeps his bright eye on us is wearing my life slowly away. He was down the day after he placed the order to see how the castings had come out and they weren't even in the sand yet.

Mother is all better. She probably worked too hard getting ready for her party. Anyway she is feeling festive now for she is having a new gown made to wear to the Land League Ball to-morrow night and she tells me it is quite stunning. I'll not go to the Ball, can have a good deal more fun at the gymnasium. You ought to see me do the traveling rings now.

Gertrude Crane has written me that we are very much alike. You and I, I mean. That is one of the most flattering things that can possibly be said about you and you want to remember it. It will come handy some time to cheer your hours of gloom when you have been indulging in mince pie or a late supper.

Lovingly,
Kate[16]

The day before, Jim had written William:

Dear Father,
I received a letter from Kate this week, and she was telling me of the new device you got up for your lathe aprons. She didn't make it very clear as to how you succeeded in making it impossible to throw in both feeds at once, and do it without extra expense. News from the shop makes me impatient to be at work again; but the College year is more than half over and it will not be long before I am at it again. We have a two weeks vacation beginning about March 17th. Do you want me to come home and go in the shop or stay here?

Your son,
James E. Gleason[17]

On February 12, Kate conveyed William's answer:

Rochester, N.Y., Feb'y 12th. 1889

Dear Jim,
Father is intending to write you to-day to explain his latest magnificent improvement in machine tools and to tell you that you are to come home for the spring vacation. The business continues to be more than we can look after and I am blessed if I know how we get it for every body else in town is dull as a hoe and the price of pig iron has gone away down which is a sign that business through the country is dull.

Lovingly yours,
Kate[18]

Rochester, N.Y., Feb'y 26th

Dear Jim,
Gertrude Nugent informed me that she had sent you an invitation for her soiree in hopes it would accelerate your arrival here. I will not honor her festivities for I have reached that advanced age where I don't give a picayune for parties and henceforward will only attend when I have a guest to entertain or as a very special favor. So Wednesday night will see me trying to

break the record in standing high jump at the gymnasium instead of whirling about in the mazy dance at Gertrude's.[19]

By March, Kate was eager to get on the road again.

Rochester, N.Y., March 6th

My Dear Boy,

I am trying to induce my paternal relative to let me go to Mount Vernon, O. after an order for an 84' x 84' that a concern down there are inquiring for but he does'nt give me any encouragement and if the firm fails to capture that order you will know the reason why. If the country is in a walkable condition when you arrive so that we will have an opportunity to reach the neighborhood where rail fences abound, I will be able to show you some things that can be done in the line of vaulting by a finished gymnast. I am so far ahead of everybody else I know in that line that I prance about like Goliath among the Philistines.

Very lovingly,
Kate[20]

Rochester, N.Y., March 9th

Dear Jim,

The above date is wrong, it should be the 10th, the Holy Sabbath Day and as I have just written a very large pile of business letters and it is along about supper time and I am correspondingly hungry, you will not be honored with my attention very long.

Mr. Walker of the Walker Manf. Co. Cleveland was here all day yesterday and I found him an uncommonly interesting man. He seems to be quite sure that the gap that will be left in the world when he is taken off will be a permanent vacancy and he is such an undoubtedly brainy man and good mechanic that one is tempted to agree with him in that.

Business shows no signs of slacking up for which we are devoutly grateful.

Lovingly,
Kate[21]

Kate did not have the same fond feelings for Professor Robert Henry Thurston, director of the Sibley College of Engineering, that she did for Professor Albert S. Smith. Thurston, first president of the American Society of Mechanical Engineers (ASME), had been director of Sibley College when Kate returned there in January 1888 and was known for his "self-conscious formality."[22] Kate made no bones about her disdain for the esteemed director.

Rochester, N.Y., April 4th

Dear Jim,

Let me know whether you will be allowed to take the machine designing or not. And if you can't what [you will] take in its place. I would advise you to ask Prof. Smith's advice for the good it will do you and then you might ask Prof Thurston's for the good it will do him.

Lovingly,
Kate[23]

Five days later, she wrote:

Dear Jim,

When I read in your letter that you had called on Gertrude [Crane] and the very next day spent the afternoon on the lake with her, I felt tempted to remind you that you have a family depending on you for support for some few years to come and that you can't afford therefore to fall in love with Gertrude but then I thought of Jo's pea story and I concluded I wouldn't. The pea story goes like this,—the mother went off and locked the children up in the house for the day and before she went, she told them to be good children and not put any peas up their noses. They hadn't thought of that special piece of mischief until she suggested it so when she came home she found that the children had all put peas in their respective noses and as a consequence they had to be amputated. People at Cornell get remarkably spoony in the spring-time, not that I have ever experienced it but time and again I have seen how others were afflicted.

Was just trying to think if there wasn't some important news around the shop to tell you when it burst upon me that you had not yet heard about the foundry dividend. We have been making money at the rate of 12 & 1/2 per cent per annum for the first six months. What do you think of that? What will you give me for my stock? We are keeping this tremendous profit a dead secret for fear all the capitalists in town will embark in the fdy. business and ruin our trade.

They are having a mission for men at the cathedral now and last night one old sinner got so uproarious that the patrol wagon had to be summoned. The patrol wagon backing up to the cathedral didn't look exactly proper. He wasn't any of our friends.

Lovingly,
Kate[24]

NOTES

1   William Gleason to James Gleason, November 4, 1888.

2   James Gleason to Kate Gleason, November 5, 1888.

3   Ibid., November 6, 1888.

4   Kate Gleason to James Gleason, November 8, 1888.

5   Ibid., November 10, 1888.

6   Ibid., November 14, 1888.

7   Charles Warner to James Gleason, November 14, 1888.

8   Variations in the spelling of names occur throughout Gleason correspondence: Gleason is sometimes spelled as Gleeson or Gleson; McDermot is sometimes spelled with two "t's"; Eleanor is sometimes spelled "Elinor"; Emmet is sometimes spelled "Emmett"; etc.

9   Kate Gleason to James Gleason, November 21, 1888.

10 Ibid., November 26, 1888.

11 Ibid., December 7, 1888.

12 James Gleason to Kate Gleason, December 1888.

13 Kate Gleason to James Gleason, January 6, 1889.

14 Ibid., January 9, 1889.

15 Ibid., January 30, 1889.

16 Ibid., February 5, 1889.

17 James Gleason to William Gleason, February 4, 1889.

18 Kate Gleason to James Gleason, February 12, 1889.

19 Ibid., February 26, 1889.

20 Ibid., March 6, 1889.

21 Ibid., March 10, 1889.

22 Sinclair, A Centennial History, 35.

23 Kate Gleason to James Gleason, April 4, 1889.

24 Ibid., April 9, 1889.

CHAPTER 5

Spreading Wings

Rochester, N.Y., April 18th

Dear Jim,

Here is a chance to distinguish yourself. The Empire Well Auger Co. of Ithaca inquire[d] this morning for a 30' taper attachment lathe to turn 5' 6" between centers. And we have written them that our Mr. James Gleason is in their city and we have requested him to call. Enclosed please find list price of lathes about that size. We are quoting sharp up to list now because we are so busy but as we are particularly anxious that you should take the first order you go after same as I did, you are to quote them 5 per cent off this list and if you are crowded 7 & ½ per cent. Time of delivery six weeks. Though we are quoting ten weeks to three months now but Father wants you to take this order badly, so do I. If they say anything about Lodge-Davis's lathe you might casually mention that Kesselman & Co. of Butler, Pa. has just put out a L.D. Co. taper attachment lathe and put in one of ours but don't say it unless you have to because it is poor policy to run down other people's tools. We will make the 26" lathe swing 28" without extra cost and the 30" swing 32". I enclose circular giving dimensions, a photo of the 30" lathe and a photo of the new shafting lathe because it is the only photo we have yet showing the new attachment for reversing feed in the apron. You want to explain that to them and tell them we are very anxious to have one of our latest improved lathes in the town.

Wear your best clothes and your sweetest smile and get that order if it's to be got.

Yours lovingly,
Kate[1]

Apparently, Jim did not get the order, and on April 22, Kate wrote, some-what soothingly:

Dear Jim,
Yours rec'd. Once in a while the Gleason family does get left and there's nothing like selecting an appropriate time for that little ceremony. Now we didn't want that auger Co. order any more than a cod-fish wants a gold watch and chain, it takes all my leisure moments now trotting up street to wire belligerent customers that we are working all hands on their particular tools and will ship them early next week or next month or something to that effect and if I had one more on the list it would probably be the last strain that would make my gigantic intellect a blank.

Enclosed you will please find a token of my affection in the line of a neck-tie. Please return the photo of 30" lathe next time you write, not the shafting lathe, believe you said you were going to honor Prof. Canaga with that, but the regular. These cost 12 & 1/2 cts each and I can't afford to waste it on you now your customer has gone up in smoke.

Lovingly,
Kate[2]

A week later, Jim replied, deftly changing the subject:

April 28, 89
Dear Kate,
Yours received with necktie all safe: Please accept quite a number of compli-ments on your ability to select neckties of the latest. Now Kate you are to receive an invitation to the D.U. reception for May 10. If Emma Michel and yourself are in the mood for a spring spree it would be a grand idea to accept. Emma will get one also.

Yours,
Jim[3]

By this time it was Andrew's turn to forego his high school studies and to work in the shop with his father and sister. On April 28, 1889, he reported to Jim at Cornell:

There has been six lathe hands layed off during the last month; three were sick and three had their fingers taken off. I have been particularly lucky. I haven't got cut since I began. You see it's got to a place where my work makes a big difference.

We are going to have chicken and greens and lemon pie for supper. Oh don't you wish you were home. Emma is sick today so this will be the first Sunday since you went away that she has not been over to supper.

A.G.[4]

On April 30, Kate declined the D.U. invitation:

Dear Jim,
I have my fears that your whole chapter will be plunged in gloom at the receipt of regret cards from Miss Michel and myself.

This year Andrew's name appears in the directory for the first time 'Andrew C. Gleason, Machinist, 10 Brown's Race, [boards] 104 Platt.' He has ordered a special copy for his own use so that he can look at it every night and morning.

... I went to the Ignorance Club long enough to see Dr. Dolley again and have everybody tell me how much they were missing me.[5] Haven't had time to go to the club but a few times this winter. Then I put in an hour at the gymnasium where I distinguished myself in the face of an audience of at least one hundred by putting Miss Harris and her brand new suit on the floor in a particularly dusty spot and sitting on her until she had taken back some disparaging remarks she made to me. And I wound up the festivities by an hour at the Turkish bath rooms.

Lovingly,
Kate[6]

On May 19th Jim wrote Kate telling her that he would be home in three weeks and asking whether she had any idea of what sort of work he would be doing when he got there. He also asked if Andrew was doing any preparing, or whether he was putting all of his time in the shop.

Rochester, N.Y., May 22, 1889

Dear Jim,
Sunday we inaugurated the family picnics for the season. We drove to the bay and we met there Emma's latest victim, Mr. Newitt, and a young man named Cortez, descendant of the celebrated conqueror of Mexico; the family at present though does not own Mexico, at least this Mr. Cortez doesn't. As far as I could see Sunday, his possessions were limited to a vague, sweet smile and three 2 oz. fish.

Lovingly,
Kate[7]

In the summer of 1889, Jim left Cornell and returned to Rochester, where, he confided to some of his fraternity brothers, he expected to stay on in the family business. Word spread, and many of his friends wrote to 104 Platt Street, encouraging Jim to reconsider and stay at Cornell. There is no doubt that he and the Gleasons were conflicted. Jim was a great help to his father, but the family knew that he would be even more valuable in the future after further studies in mechanical engineering. In mid-September, the family decided to send Jim back to Cornell to continue his education.

Two months later, however—on November 25, his twenty-first birth-day—Jim was suddenly called home. That morning, the Gleasons' Genesee Foundry building had burst into flames. By the time the fire department arrived at the scene, the whole structure was blazing. The roof collapsed, dropping the building's three floors and all its heavy machinery into the cellar. William's machine shop, located in front of the foundry, was also on fire, but that building proved to be sturdier and withstood the flames. Fortunately, no one died in the inferno, and though the Genesee Foundry was a total loss, William's shop was repairable, and some of his machines survived the disaster.

Kate, who had just celebrated her twenty-fourth birthday the day be-fore, had been at the scene all morning. According to newspaper reports, "She took the affair good humouredly, remarking that 'it was a nice way to celebrate Thanksgiving.' Her first care was to secure the insurance policies, which she kept in her own hands."[8]

The inferno, Jim later recalled, was the most memorable experience of his life. "I went out to the ruins," he remembered, "and started pulling parts out of the debris and came across Father, hands in pockets, walking around as though in a daze. Father said, 'Who sent for you? I don't need any help around here.'"[9]

Although insurance helped offset expenses, it did nothing to ease the loss of all of William's patterns and drawings. At the age of fifty-three, he faced the challenge of starting all over again. Despite the trauma and loss, however, the destruction, William later admitted, was "the finest thing that ever hap-pened," because it spurred him to design improved machines and new inno-vations.[10] "To another man," Eleanor recalled, "it would have meant the end." To William, however, "it meant simply greater endeavor. With the fiery ener-gy that was so characteristic of him he started next day to erect single handed a new and better business from the ruins of the old. First he discarded all his old designs and in less than three months had produced with his own hands drawings for a complete new line of lathes and planers."[11]

The next year, William sold a gear planer to a company in England and traveled to Newcastle-on-Tyne to install it at the Armstrong Whitworth & Co. factory. He journeyed to Ireland on the trip, likely his first visit to the "Old Sod" since leaving Tipperary forty-two years earlier. Jim, meanwhile, returned to Cornell and continued his studies through the spring before entering the Gleason business full time.

By the end of 1890, Kate, now twenty-five, was secretary-treasurer of the machine shop, which the family had reorganized as the Gleason Tool Company. She owned 26 shares in the Genesee Foundry Company and 102 shares in the Gleason Tool Company, while William owned 107 shares and 878 shares in the firms, respectively. Kate was also continuing her education, attending night classes in mechanical engineering at the Mechanics Institute in Rochester.[12]

The business, happily, was recovering, but Ellen Gleason was starting to lose her health. A tiny woman, spirited and energetic, Ellen had a deep-rooted sense of equality and justice and was an exceptionally fine horse-woman, keen on racing and able to handle a recalcitrant steed better than anyone in the family. But she had stopped eating, a problem that her doctor told Kate was partly mental; in fact, Ellen was suffering from an undiagnosed case of tuberculosis.

In the winter of 1891–1892, seeking a warmer climate, she and Jim traveled to Sacramento, California, to visit her older sister, Catherine (Kate) McDermott Hughes. Kate Hughes and her husband, Owen, had ten children, all living under their roof, and they warmly welcomed their Rochester relatives. Jim wrote home in February, "The climate in Sacramento is something grand, just like a day in May at home. The air is bracing which appears to be just what Mother wants. Her voice sounds as clear as a bell and what is more she eats here things she hasn't tried before in a long time. An egg in the morning, milk and toast. Aunt Kate says she is getting better. The fact is Mother is as happy as the day is long, mending the boy's clothes, occasionally doing some cooking and the rest of the time standing me on my head."[13]

Ellen's improvement, however, was short-lived. She had insisted on accompanying Jim to San Francisco, where he visited machine shops and searched for new markets for Gleason tools. The city's damp, foggy weather and chilling winds took a toll on her health. Ellen, Jim wrote his sister, had grown very weak and frightened.

Aunt Kate was afraid that we might have trouble getting her back to Sac., all the effect of not being able to eat. She realized the fact that she was sinking. Then it was that she consented to see this Doctor Huntington and that terrible week in S.F. has saved her. Now she minds the doctor as she never would before. This doctor gave her a thorough examination and told her why she couldn't eat and made her able to eat moderately in four days and has made a decided advance in her whole make up in two weeks and a little over. Aunt tells me that the medicine Mother takes has only been known to the profession for about a year. The Doctor Huntington is the S.P.R.R. [Southern Pacific Railroad] Company physician and a fancy priced man.[14]

Ellen, however, remained ill through the spring of 1892, and the doctor urged her to stay in Sacramento until summer. Jim planned to return to Rochester, via Salt Lake City and Denver, in early April, while his mother would travel east in July, accompanied by her niece Emma Hughes. "Mother appreciates her case now," Jim reported, "and is anxious to do anything that will help towards her recovery. She will visit a number of springs and health resorts in this vicinity."[15] But Ellen was "a trifle stubborn" and impatient to be at home; against all entreaties, she journeyed back to Rochester with Jim in April.

Family finances, like Ellen's health, were deteriorating badly. Another panic that spring had triggered a financial crisis. Business, according to Kate, was "about the worst in the world. There were alternate periods of inflation and depression," Kate said, "that were utterly beyond our control. We made just as good tools at any time, but in boom years when orders rolled in we had to extend our plant facilities to get out the work, and to do this we borrowed money. Then bad times came, and we found ourselves with notes due and no work to bring in money to pay them."[16]

By the fall, Kate, in charge of selling and finance, did her best to get orders, with some success, "but no orders could save us," she said, "unless I could induce the banks to accept customers' notes to pay our notes. As I tried to persuade bank officers to extend our credits I began to appreciate what feminine charm might do. I thought of all of the wiles of all the great adventuresses and wished I could learn some to use on the bank executives."[17]

There was, however, something of a silver lining. Andrew was now studying at Cornell, and with business slow, William and Jim could spend more time improving their machines. That year, Jim enhanced his father's original bevel gear cutter by automating it, and both Kate and Jim encouraged their father to concentrate on gear-cutting machines instead of the general machine tools he had been manufacturing, a transition that the company gradually began to make. It was a pivotal decision, since it coincided with the dawn of the automobile industry, which relied on gear-cutting machines. In 1886, Karl Benz, in Germany, had received a patent for the first gasoline-fueled automobile, and by 1893 Charles and Frank Duryea had demonstrated the first gas-powered vehicle in the United States. Gear-cutting equipment would eventually be essential to the fledgling industry.

That year, Kate was distressed about blanks in the order book, and she was feeling ill. Her doctor suggested a vacation in Atlantic City, but that didn't suit Kate, since she knew no companies in the town that needed Gleason products. Instead, she hatched a better idea. Europe was not yet experiencing the same business downturn as the United States, so Kate decided she could relax on an ocean voyage and visit companies abroad. William agreed to the bold proposal. He knew that arguing with Kate was a losing proposition, and he had enough confidence in her ability to be successful. Others, however, were shocked that Kate, at twenty-eight, would embark alone on an overseas business trip. "Three days before I sailed," she recalled,

> I mentioned my going to a salesman from Brown and Sharpe. He was aghast. "But—do you know anyone there?" he asked. I said I did not. I sold to people in America whom I did not know, so why not abroad? He shook his head, went back and told his firm, and they sent me a sheaf of letters of introduction. This firm was larger than ours, and in the same line. On my return, knowing what those letters had done for me, I sent them a check as commission on my orders. They returned it to me. That is one of the many times men have gone out of their way to lend me a helping hand.[18]

She packed lightly (just one black cashmere dress, the extent of her presentable wardrobe), gathered the two hundred dollars the trip would cost her, and embarked from Montreal on November 20, boarding the cattle steamer Mongolian for a two-month voyage across the Atlantic Ocean. She was the only woman passenger among fourteen men, a delightful experience, Kate recalled. The gentlemen took turns walking the deck with her, using a stopwatch to ensure fairness. "Naturally, I was feeling pretty good," Kate remembered.

> But the purser spoiled it all for me. He told a story of a time when, after six months in the Congo without sight of a white woman, his major's wife got on the return steamer and he found her beautiful beyond words. But a few days later some young English girls boarded the ship, and he suddenly saw that the major's wife wore a wig and had false teeth. The moral was obvious.

I said good-by to my fourteen ship admirers, got my letters, and prepared to work for orders. I was fortunate enough to bring some back from the best firms in Scotland, England, France and Germany, and that was the beginning of a real foreign business for us.[19]

Kate returned to Rochester from her first overseas trip with a new perspective on the world and her place in it. She was a pioneer in establishing overseas markets for products manufactured in the United States, and she had a new sense of her own powers. Kate had already established herself, a journalist wrote, as "a witty, knowledgeable and spunky non-conformist and nobody's fool when it came to machine tools, but the trip abroad brought her even more self-confidence and a polish which would stand her in good stead the rest of her life."[20]

Before the trip, Kate had neglected her looks. One of her former teachers and even Susan B. Anthony had chided her about her appearance, and more than one person had told Kate that she smelled like horses, which she loved to ride.[21] Now, she realized, after two months in one black cashmere dress, it was time for a change. "Shortly after the trip," Kate recalled, "I began to consider my clothes. Apparently, everybody else had been considering them for some time; but to me dress had seemed of little importance."[22] Now, she "went in for extremely feminine attire. I had my hair dressed and wore violets in my muff, and had some soft, frivolous gowns made. This attention to dress repaid me well. Some of my customers spoke to me twenty years after about a certain dress or hat that I wore when I made a sale. I learned to value clothes, to love clothes, and to use clothes."[23]

Kate transformed herself into a handsome, charming, gracious woman who conveyed the impression of robust vitality, acuity, wittiness, and immense power. Although her height, by today's standards, was slightly below average, as was that of her siblings, she was variously described as being of average height or tall. Perhaps the high-heeled shoes she wore made her seem taller, and her erect posture and commanding presence added to that impression. Her sparkling and expressive blue eyes helped capture the attention of her listeners, though she had a "cast" in her left eye, a weak muscle that sometimes caused it to lag the right eye in focusing; it may not have been noticeable in person, but it was caught in photographs. Kate also tended naturally toward stoutness, and she battled the bulge throughout her life, dieting on and off and keeping records of her weight. In general, after her initial trip to Europe, she paid great attention to her appearance, changing her hair and clothing styles to keep up with the times. Despite her many attractive physical qualities, however, it was Kate's personality and marketing skill that made the biggest impression. On those rare occasions when self-confidence failed her, her bravado trumpeted her presence and power.

She and Jim were now spending a great deal of time on the road pursuing orders for machines and establishing the firm's position as widely as possible. Kate found she had little trouble getting her foot in the door. "Most people," she explained,

> were curious to see me. Susan B. Anthony, who was a great friend of my mother, had impressed one fact upon me while I was growing up: "Any advertising is good. Get praise if possible, blame if you have to. But never stop being talked about." I have come to believe that absolutely. In those early days I was a freak; I talked of gears when a woman was not supposed to know what a gear was. It did me much good. For, no matter how much men disapproved of me, they were at least interested in seeing me, one distinct advantage I had over the ordinary salesman. I dealt wholly with men—no women were then running factories and foundries.[24]

There were limits, however. Salesmen in those days entertained their customers, much as they do today, at lunch, dinner, theater, sporting events, and parties at less-than-upright establishments. "By all the known laws of my day," Kate stated, "I could not do this"; she operated in an all-male profession and was at a distinct disadvantage, according to Victorian conventions.[25] But Kate was determined to find ways to entertain her customers. She had always had a powerful facility for language, as her letters attest, that was an invaluable sales tool. Kate blended this natural eloquence with an acute sense of the comic, becoming a first-rate raconteur who could "spin a marvelous yarn," according to Jim. "Many a time," he declared, "I've heard Father give one snort at a yarn of Kate's, and then give way to helpless laughter."[26] She even took voice lessons and learned to put her full, low-pitched voice where she wanted it to be, just as she put herself in business environments where other women were not likely to be.[27] With her positive outlook, abilities, and determination, she became known as a charming and formidable business competitor.

Soon, Kate was also shouldering more responsibilities within the family. Her mother's health had continued to deteriorate in the harsh Rochester weather, and as the winter of 1895 approached, it became clear that she would have to move to a drier climate. The family settled on San Antonio, Texas, and William, Ellen, and Kate's fourteen-year-old sister, Eleanor, departed by train, stopping briefly in New Orleans to visit Ellen's brother and his family.

It was a miserable journey, William wrote Kate from Cincinnati on their way south. The cars were crowded, and Ellen was having a hard time breathing, but it was a cold day, and other passengers objected when they tried to open the window. The obliging conductor finally allowed Ellen to sit on the platform in the open air, where she could breathe more comfortably, though it was against the rules. To make Ellen more comfortable, William wanted to secure a stateroom for her on the next leg of their trip, but he knew she would object to the extravagance. "She is stubborn, nothing new, but more than usual on account of being tired," he wrote Kate.

> She will kick at the expense. She almost created a scene at the hotel this evening because the two rooms I secured was according to her idea of economy too grand. One was a parlor with a bed and sofa. I designed the sofa for Eleanor and a small room adjoining for myself, all of which I considered the neatest kind of economy. But my efforts were not appreciated. She wanted two small rooms and have Eleanor sleep with her. It come my turn to get mad and I found it worked all O.K.[28]

On October 8, William wrote again from San Antonio, where they had enrolled Eleanor in school. The weather was better, and Ellen had gained three pounds. Still, she weighed a mere eighty-eight pounds, a hundred pounds less than her husband. Although she had contracted a cough in New Orleans, it was milder than the coughing fits that wracked her body in Rochester. "I have not done any worrying about the shop but I done some about Ellen," William acknowledged. "I was not sure that we were going to get here. I could only guess her condition as she carefully concealed it but now she admits that she herself did not expect to reach here alive."[29]

By the middle of the month, Ellen had improved enough to allow William to return to Rochester. Once home, however, he was quickly summoned back to San Antonio. "It seems as I get away from the cares of the business," he wrote in an undated letter from Chattanooga, "the full force of the calamity that we are threatened with is revealed in full, and I have put in two very hard and sleepless nights."[30] Months later, on February 26, 1896, Ellen McDermott Gleason, at age fifty-one, died of consumption in her home on Platt Street. Her funeral was held two days later at St. Patrick's Cathedral in Rochester, and Kate, at age thirty-one, slipped naturally into the role of mater familia.

NOTES

1   Kate Gleason to James Gleason, April 18, 1889.
2   Ibid., April 22, 1889.
3   James Gleason to Kate Gleason, April 28, 1889.
4   Andrew Gleason to James Gleason, April 28, 1889.
5   Dr. Sarah Dolley, the first woman to practice medicine in Rochester, founded the Ignorance Club in 1880.
6   Kate Gleason to James Gleason, April 30, 1889.
7   Ibid., May 22, 1889.
8   "Early Morning Fire: William Gleason's Machine Shop and the Genesee Foundry Burned," *Rochester Union and Advertiser,* November 25, 1889, 6:6.
9   "James E. Gleason Acclaimed," *Rochester Democrat and Chronicle*, October 7, 1945: B, 6.
10   Eleanor Gleason, speech, January 29, 1931.
11   Ibid.
12   Kate attended classes from 1889 to 1891.
13   James Gleason to family, February 1892.
14   James Gleason to Kate Gleason, March 14, 1892.
15   Ibid., March 29, 1892.
16   Bennett, "Kate Gleason's Adventures," 169.
17   Ibid., 171.
18   Ibid.
19   Ibid.
20   Bartels, "First Lady," 11.
21   Ibid., 12.
22   Bennett, "Kate Gleason's Adventures," 171.
23   Ibid., 171–72.
24   Ibid., 170.
25   Ibid.
26   Ibid., 173.
27   Kate cultivated her natural vocal talents by taking voice lessons at the Mechanics Institute (later named Rochester Institute of Technology) from 1907 to 1908.
28   William Gleason to Kate Gleason, from the Grand Hotel in Cincinnati, Ohio, September 28, 1895.
29   Ibid., from San Antonio, Texas, October 8, 1895.
30   Ibid., from aboard the Southwestern Limited train in Chattanooga, Tennessee, n.d.

CHAPTER 6

Myths and Realities

B Y 1898, THE GLEASON TOOL COMPANY was focusing exclusively on bevel gears, which were used in the flourishing bicycle business. Some bicycle manufacturers were now starting to produce automobiles that incorporated bevel gears as well as other bicycle parts. As a result, the Gleason Tool Company was expanding rapidly, with solid enough prospects to erect a new brick building. With Kate, Jim, and Andrew all working effectively, William decided to follow in Kate's footsteps and travel to Europe to relax and establish new markets for Gleason machinery.

After visiting Ireland, he spent two days in London before visiting Manchester, England. "It is very lonesome over here so far from home," he wrote Kate on October 28,[1] but four days later, William was feeling more positive about the trip. On November 1, he wrote,

> I have been to Hollingsworth's this afternoon. Churchill has two very bright men here located in Manchester, both machinists, but never seen a gear planer in motion until this foor noon. What we need is a good man here that could go in too the shop and run the machine. If Andrew could be spared there is a big field here but we can talk that over later. I could stay here indefinitely and find something new every day. The heads off departments don't get to work before 10 and generally take lunch at 1 pm. When they frequently don't turn up after that event but they generally dispatch a good deal of business in a short time. They don't know anything about mfg.[2]

From Glasgow, on November 7, he reported that "our gear planers are becoming well known throughout the country and I believe quite a number of them will be sold the comeing year if business keeps on as brisk as it

51

is at present."[3] Three days later, William was in Berlin, where he wrote to Kate on November 11:

> I arrived here late last night and you can't have the least idea off the plight I was inn. I could not make myself understood. This city seems to be a festive lot and iff Paris beats it I will be surprised. It is a fine city and I like the appearance off it mutch better than London or Glasgow. If I go to Vienna it will delay me at least one weeck but I think I had better go as they seem very anxious to have me.[4]

It was a successful trip. The Gleason firm sold $10,135 worth of machines to Germany and Austria and another $18,250 worth to the rest of Europe. By the end of 1898, the firm's capital was $128,700, and the profit for the year was $10,400.

The next year, after taking a cruise in the Gulf of Mexico and visiting her Uncle Andrew (her mother's brother) and his family in New Orleans, Kate celebrated her brother Jim's wedding to Miriam Blakeney, on October 18, 1899. A trim, brown-haired young Rochester native whose interests ran to gardens, music, and the Presbyterian Church, Miriam had caught Jim's eye when she was working in the Gleason Company. Their marriage, however, was unhappy from the beginning. In later years, the strain was public. Miriam and Jim maintained separate lives while sharing the same home. They rarely ate together, often arriving separately at the Genesee Valley Club for dinner, and dining at separate tables. At home, the local grocery store would deliver food to Jim's cook and, later the same day, would make a separate delivery to Miriam's. The family considered Miriam an exceedingly cold woman and blamed her for the problems in the marriage, but neither she nor Jim seemed to have strayed in search of affection, and they remained married for sixty-two years.

Eleanor, meanwhile, was blossoming into a beautiful young woman. She had many beaus, but Kate kept a tight rein on her, allowing her to go out no more than once a week and warning her about men who would marry her for money. Under Kate's protective custody, Eleanor became painfully shy and nauseous before she went out on dates. The specter of Kate giving her beaus the evil eye surely had a withering effect on any budding romance. With the best intentions, Kate laid the groundwork for Eleanor's lifelong spinsterhood. She was an independent-minded and capable person, but she never managed to escape Kate's powerful orbit.

In the fall of 1899, Eleanor enrolled in Cornell University, although she chose not to follow in her siblings' footsteps. She had no interest in engineering, preferring literature, and there was little pressure on her to enter the family business. Still, Eleanor was a pioneer in her own right. The year she began college, a debate was raging at the University of Rochester over the admittance of women, who were allowed to take classes as "visitors" but denied enrollment. Kate's Ignorance Club had been agitating for coeducation, and in 1898 the university's trustees finally agreed. They required, however, that proponents, led by Susan B. Anthony, donate a hundred thousand dollars to help fund coeducation at the institution. Two years later, facing commitments of only forty thousand dollars, the trustees reduced their demand by half, and Susan B. Anthony personally pledged her life insurance to cover the last two thousand dollars toward the goal.[5] Eleanor decided to transfer from Cornell and was one of thirty-three women, including five other transfers, who enrolled in the University of Rochester in September 1900. When she graduated in 1903, she was one of the first of her sex to earn a degree from the university as well as the first in her family—and the only one in her generation—to graduate from college.

Kate was breaking boundaries, too. In 1900 the Gleasons exhibited their gear planing machines at the Paris Exposition, a time-consuming and costly undertaking with potential for widening the firm's customer base. When they learned that the space allocated to the Gleason Tool Company was very dark, and under a staircase, Kate decided to take matters into her own hands. She left immediately for Europe, traveling aboard the Statendam in early April. Arriving in Paris, she discovered that a dozen firms were also poorly situated, including a Swedish firm that had appealed for help to Sweden's minister to France. Kate relied on her own brand of diplomacy. "I slipped back to my hotel," she recalled, "put on my prettiest frock, and the most feminine, laciest hat you ever saw, and went boldly to the general manager's office."[6] Gone was the black cashmere dress, along with any lingering thought that being a woman was a drawback in business. Kate was shown directly in to meet with the exposition manager, past a waiting room full of people eager to see him. She encouraged him to talk and "listened long to his tale of woe and condoled with him all along. I asked nothing," Kate recalled. "In the end he inquired about my exhibit, and when I mentioned its location he rose at once, took me out and offered me a choice of nine of the best places in the building. The other American exhibitors kindly helped me to move, and the remaining denizens of the dark corner watched open-mouthed."[7]

Kate knew that she had staged this coup in front of a gilt-edged audience of machinery manufacturers from around the world, and her reputation in the machine-tool industry, already formidable, was spreading around the globe. As a business acquaintance put it, Kate Gleason "was an attractive young lady who had broken several of the Victorian concepts about woman's place being in the home." With her "flair for the mechanical," she pursued "a career that had hitherto been considered the exclusive bailiwick of men with long beards. Instead of sitting at home tatting or making samplers, Kate spent her youth learning her father's business from the ground up, both in the shop and in the field" and "knew as much as any man in the business."[8]

She was not shy about tooting her own horn, mindful of Susan B. Anthony's advice, but all too often the stories about Kate were exaggerated to the point of fiction. Years later, she told journalist Helen Christine Bennett about visiting a customer and noticing a squeak in one of his Gleason planers. Kate diagnosed the problem and had her father send out the needed replacement. "The yarn that arose from that simple act," Kate said,

> thrived and grew into a wonderful tale. Father had visited the shop a short time before I did. He also noticed the squeak, but as it did not impair the work forgot to send the new fingers until I reminded him. The owner of the mill began spreading the story of how Old Man Gleason himself, the inventor of the machine, had been there and never fixed that noise, and then his girl came along and fixed it in two minutes.[9]

In subsequent iterations, Kate was wearing overalls and carrying a tool bag, something she adamantly denied. "I never have worn overalls and never will wear them, but that tale spread all over the country."

Other stories about Kate persist to the present day. Henry D. Sharpe, Jr. of the Brown and Sharpe Company recalled a story his father often told about a visit Kate made to their Rhode Island manufacturing plant. The managers gave her the grand tour of everything in the facility except the foundry, since her escorts were unwilling to expose her to the rough and tumble of a masculine enclave, with workers occasionally stripped to the waist. Kate would have none of it, however, and insisted on visiting the foundry. With much consternation, the men formed a phalanx around her and hurried her through the facility so that she would not be able to get a good look at anything. Refusing to be railroaded, Kate spied a large mound of pig iron in the center of the room, broke away from her guardsmen, unceremoniously mounted the pile, and observed the foundry operation in full from her towering perch. The Brown and Sharpe men were horrified, and another colorful tale about the indomitable Kate Gleason had its beginning.

None other than Henry Ford spread the most famously erroneous tale about Kate and her talents. Ford referred to the Gleason gear planer as "the most remarkable machine work ever done by a woman."[10] Even though he was corrected more than once, he persisted in attributing the invention to Kate. He also reportedly said that "history is more or less bunk," and in this instance, he was largely correct.[11] Kate explained to Bennett how unfair that story had been to her father, the actual inventor, as well as her brothers, who had improved the revolutionary machine. Ford never retracted the comment, however, and it became part of the lore of the machine tool business.

Kate's salesmanship had a life of its own, and she was never able to get the genie back in the bottle. On May 18, 1910, however, she wrote a letter to the New York Times editorial department, attempting to correct the record:

> There was a paragraph about me in your paper of May 15th, that I wish were all so,—but it isn't. The paragraph is headed 'Feminine Mechanical Genius' and credits me with designing our Bevel Gear Planer. The nearest I have come to designing it is in having a Father and brother smart enough to do it. My place in the business is Secretary and Treasurer. You see I have captured two jobs but neither of them have anything at all to do with designing.
>
> Is there any chance that you can overtake that mistake with a correction. It will oblige me very much if you will try. About the most important training for the treasurer is not to take what does not belong to one and it looks to me as though I would be falling down on my own job if I get credit for other peoples work.
>
> Yours truly,
Kate Gleason[12]

Not all of the publicity Kate attracted, however, was undeserved. As Fred H. Colvin noted in his book, *60 Years with Men and Machines*:

> It was through A. H. Carpenter that I first heard about one of the most interesting women I have ever known ... This woman, a kind of Madame Curie of machine tools, was the renowned Kate Gleason, daughter of William Gleason, the founder of the Gleason Works of Rochester, New York, and at that time the only woman in the world connected with the machinery industry in a major sense ...

Carpenter told me the story of how Kate came one day to the Acme Machinery Company, of which he was the manager, intent on selling them a batch of Gleason gear cutters. Carpenter was certain that this slim young girl knew little or nothing about the product she was selling and determined to prove this at least to his own satisfaction by calling in Thompson, his gear expert, to test her knowledge.[13]

As Colvin recounts, Thompson asked Kate to explain why her father's machines might be superior to the ones that were in the Acme shop, warning her of his expertise in gears and gear-cutting machines. "I am very glad to hear that, sir," Kate replied, "for it will make what I have to say that much easier." She then proceeded to sketch in the air with her gloved hand the intricacies of the Gleason Bevel Gear Cutter, using engineering terminology and displaying an understanding of gear technology that left Thompson bewildered. He attempted to bluff his way through, but she had lost him somewhere between a spiral bevel gear and helical teeth. The uncomfortable Mr. Thompson found an excuse for a hasty exit, and though Kate sold no machines that day, she had the respect of Carpenter from that time forward.

Kate herself knew that the stories about her were her greatest asset and inflated her contributions to the family business. "Before I began work," she recalled,

> I was often in the society of my mother's friends, who were ardent suffragists, and some of them warned me that men would be my business enemies. I went into a man's field, that of machine-tool making, but found the exact opposite. From the time I began to work men stretched forth helping hands to me; men talked to each other about me and spread my fame far and wide, and men were so eager to give me due credit that they set their imaginations working, and told yarn after yarn about things I had never done. They were a wonderful help in selling; but they were monstrously unfair to my father and brothers.[14]

Kate's salesmanship and growing reputation were good for the firm, and the first few years of the twentieth century were strong ones for the family business. The Gleason Gear Planer was essential to the automobile industry, which was in its infancy, and in June 1902 the Gleasons themselves joined the growing ranks of motorcar owners, purchasing their first auto for the scandalous price of four hundred seventy-five dollars. They promptly smashed it on Labor Day, adding another $61.20 in repairs to the total cost of the car.[15]

Four months later, on January 10, 1903, the Gleasons reorganized their company to reflect changing business conditions. They dissolved the Gleason Tool Co. and Genesee Foundry and incorporated their enterprise as the Gleason Works. Kate was again appointed secretary and treasurer, and Jim moved up to the role of general manager.

Jim's domestic responsibilities were growing, too. In July, his wife Miriam gave birth to a son, Emmet Blakeney Gleason. An only child, Emmet grew up in an unhappy home environment that was poisoned by Miriam, who constantly enumerated her grievances against her husband. A "mama's boy" who preferred his mother's indulgences to his father's scolding, Emmet grew to adulthood in a crippling miasma of hate and privilege. Although he followed his father into the family business, eventually succeeding him as president of the Gleason Works, he had, by that time, become a serious alcoholic, soothing himself with an excess of food, drink, and lavish spending. He went to great expense to try to establish a genealogical link between his mother's family and royalty, to no avail, and there was certainly no chance of such exalted ancestry on his father's side. On Emmet's frequent trips to New York City, he would ensconce himself at the St. Regis Hotel, entertaining blonde girlfriends who never had a chance of winning Miriam's approval. He never married and died at fifty-five, after suffering from the scourges of obesity, hypertensive cardiovascular disease, diabetes mellitus, Laennec's cirrhosis, and disfiguring psoriasis of the scalp and skin.

Soon after his birth, while Jim and Miriam were settling into their life as parents, Kate was setting her sights on exotic travels. In October 1903, she toured Andalusia, Spain and Tangiers, Morocco, and visited the Alhambra, a magnificent Moorish palace that had mesmerized her since she was nine years old, when she first read Edward Bulwer-Lytton's *Leila or, the Siege of Granada*. William, now in his late sixties, was traveling too. He increasingly relied on Kate, Jim, and Andrew to run the business. In 1904, when the company recorded a profit of $36,000, William traveled to Europe and the Mediterranean, stopping in Genoa, Granada, Algiers, Malta, Alexandria, Jerusalem, Athens, Naples, Rome, Paris, and Ireland.

It was a prosperous time. A few days after William embarked from New York City on his extended journey, Kate bought her very own motorcar for a thousand dollars. The next year, business for the Gleason Company was better than ever. In 1905, Jim invented the two-tool bevel gear generator, which halved the time needed to manufacture gears and produced a better-finished, quieter, and smoother rolling product at a lower cost. The innovative machine was in great demand among automobile manufacturers, who relied on bevel gears for driving wheels, and spurred the design of better, less expensive motorcars. It was so successful that the firm sold all its patterns and drawings for lathes and planers and concentrated almost exclusively on bevel gear machinery.

With the rush of business, the company's Brown's Race location could no longer handle the growing number of orders, so the Gleasons purchased land on University Avenue, on the site of the Rochester Baseball Club's old Culver Road Park. The Gleason Works erected a heavy machine shop annex and a twenty-two-thousand-square-foot foundry. William, in his European travels, had observed that cathedrals, with their vast interiors, would be ideal structures for a traveling crane. Kate sent away for architectural plans, and they chose to model their new foundry after the Cathedral at Pisa. The Gleasons took great care with the details, even adding a stringcourse for climbing vines on the exterior walls. Later, the company erected an office building modeled on the Pan-American building in Washington, D.C., setting it back from the street to provide perspective and areas for lawn and plantings. The landscaping, conceived long before it was fashionable for industrial firms to care about their curbside appearance, set it apart from most heavy industries. It has been referred to as "perhaps the handsomest factory in Rochester."[16]

Meanwhile, Eleanor, after graduating from the University of Rochester in 1903, attended Pratt Institute Library School in New York City. She became close to one of her classmates, Edith Hill, and often brought her home to Rochester during the school holidays. Edith had spent a year at Barnard College before transferring to Pratt to study library science, and after graduation she worked at the Brooklyn Public Library, at the munificent salary of eighteen dollars a month. Her frequent visits to Rochester soon sparked a romance with Eleanor's handsome brother Andrew, and on February 5, 1906, the two were married there by Gleason family friend Fr. Edward J. Hanna, who later became Archbishop of San Francisco.[17]

Edith hailed from an intellectually distinguished family in Yellow Springs, Ohio. Her mother was a direct descendant of Anne Hutchinson, the brilliant and courageously outspoken exponent of religious tolerance who was banished from the Massachusetts Bay Colony for her beliefs. Edith's father, Franklin Hill—who died when she was ten—had been associate treasurer and a faculty member of Antioch College in Yellow Springs between 1860 and 1871, before serving as curator of Princeton University's E.M. Museum of Geology and Archaeology. Franklin's uncle, Thomas Hill, had succeeded Horace Mann in 1859 as the second president of Antioch College before becoming the twentieth president of Harvard College in 1862.

Antioch had been poorly funded when the Hills were there; even Thomas Hill, its president, had to borrow money to supplement his meager salary, which the college never fully paid. Since Franklin had a lower position as well as a family to support, Edith's childhood was financially spare but rich in erudition. Edith, according to her daughter Ellen Gleason Boone, was a confident individual who had grown up on "mush, milk and intellectual arrogance."[18]

Three days before she married Andrew, the Gleasons hosted another event, Susan B. Anthony's eighty-sixth birthday celebration. Kate and her father were both dedicated advocates for women's rights and members as well as financial supporters of the National American Woman's Suffrage Association (NAWSA).[19] In 1900, a series of petitions had been read in both houses of Congress from women's suffrage associations across the country, urging voting rights for women in the new U.S. possessions of Cuba, Puerto Rico, the Philippine Islands, and Hawaii. One of the only petitions signed by men had come from eight Rochesterians, headed by William Gleason.[20] In 1903, Susan B. Anthony had personally presented Kate with the three-volume History of Woman Suffrage, which she had edited with Elizabeth Cady Stanton, Ida Husted Harper, and Matilda Joslyn Gage. Anthony inscribed Volume I of the set to "Kate Gleason, the ideal business woman of whom I dreamed fifty years ago—a worthy daughter of a noble father. May there be many such in the years to come is the wish of Yours affectionately, Susan B. Anthony. Rochester, N.Y. Dec. 2, 1903." At Christmas that same year, Harper presented Kate with the fourth volume of the history, inscribed "To Miss Kate Gleason in memory of a beautiful Sunday at the Country Club in the autumn of 1903."[21]

By the first weeks of 1906, however, Susan B. Anthony was frail and ill, and her friends planned an early birthday celebration for her on February 2. The Gleasons hosted the event at their Rochester home, and an account of the party was published in a newspaper and reproduced in the third volume of Harper's biography, the *Life and Work of Susan B. Anthony*:

> The commodious home of Mr. William and Miss Kate Gleason, was the scene last night of a brilliant reception in honor of Rochester's well-known citizen, Miss Susan B. Anthony, who on the 15th of the month will have completed her eighty-sixth year. Southern smilax and palms lent their beauty in decorating the rooms, which from eight to ten were thronged with representative people of the community. Previous to the reception the members of the Political Equality Club gathered around Miss Anthony,

exhibited to her the names of 122 women who had just been added to the roll, and then presented her a purse containing eighty-six dollars in gold. Following this there was introduced to the venerable suffragist a band of thirty High School girls who had formed a Susan B. Anthony League and pledged themselves to work for the movement to which she had devoted her life. Miss Anthony was deeply touched by this encouraging evidence of youthful interest in the cause most dear to her heart and greeted the young girls warmly ... Delightful music was furnished by an orchestra of women and refreshments were served throughout the evening.[22]

Blizzard conditions did not stop Anthony from traveling to Baltimore to attend NAWSA's convention, then on to New York for her official birthday celebration. The weather, however, may have weakened her health. Anthony took the train back to Rochester, attended by a full-time nurse, and died in her home of pneumonia and heart failure on March 13. On April 4, Kate wrote to her father, who was aboard ship, informing him of their friend's death. William had been in Panama inspecting work done on the canal, where engineers were using Gleason's twenty-foot spur gear planer to cut gears for the locks. He was on a twenty-seven day voyage north to California, aboard the Catapulca, when Anthony died. "I had not heard of Miss Anthony's death until I received the letter," he wrote his family from San Francisco on April 17, 1906, "but I was prepared for it as her condition was serious when I left. I am very sorry. She was a noble woman and I admired her. There is only about one of her kind in a centenary."[23]

After some sightseeing, William departed San Francisco by train that evening for Portland, Oregon, where Eleanor was working as a librarian. His quick departure, it turned out, may have saved his life. At 5:12 the next morning, April 18, a catastrophic earthquake struck San Francisco. The temblor, followed by raging fires, nearly leveled the city, killed more than three thousand people, and left about three hundred thousand city residents without shelter. On April 19, William wrote from Portland's Imperial Hotel:

My dear family, I arrived here this morning, having left Frisco about 9 hours before the earthquake. I have been with Elinor all day. The people here are very excited over the great calamity, many of them having near relatives living there also business interests. I hope the reports are exaggerated, if true, it is terrible.

My love to all.
Wm. Gleason[24]

During 1906, Kate was deeply immersed in building her dream house. After traveling to Spain in 1903, she had begun to think about building a home of her own modeled on a Moorish palace, with courtyards, columned arcades, and dazzling reflecting pools. Such a marvel would surely raise eyebrows in Rochester, a prospect she relished. After sending away for building plans and pictures of the Alhambra, she began to sketch some preliminary drawings and purchased rolling acreage on East Avenue, in the Rochester suburb of Brighton Heights. The property included an ideal building site on a gradual rise, with wide views of the surrounding country.

Although it seemed indecent to Kate to build a home that would cost more than fifteen thousand dollars, what she had in mind far exceeded that sum. She hired Edwin Gordon, of the architectural firm Gordon and Madden, to draw up a number of plans until she was satisfied. The intricate details of the Moorish design required craftsmen of uncommon ability, and Kate was able to find that level of skill among Rochester's growing Northern Italian population, whose members were establishing themselves in the construction trades. The craftsmen did such fine work on her home that Kate developed a life-long admiration and appreciation of the Italian people, especially those who brought artistry and dedication to their work. She was a demanding but appreciative taskmaster—"I have always placed great faith in my fellow man," she explained, "and I find that in response they give me their best"—and she grew to prefer Italian workers above all others in her construction projects.[25]

In 1908, she moved into her new home, which she described as a "royal seraglio."[26] Kate's Alhambra, named Clones after her mother's birthplace, reflected her romantic imagination. The home was constructed around a central, rectangular courtyard enclosed by a thirty-five-foot-high vaulted glass roof on ball bearings that could be opened on pleasant days. Slender, smooth columns surrounded the courtyard on three sides, under archways ornamented to the ceiling with intricate faience mosaics. As a nod to the climate in Rochester, which was far colder than in southern Spain, Kate placed heating coils between the skylight and the tiles below it. In the center of the court, planters filled with lush foliage and thirty-foot palms surrounded a shallow reflecting pool.

Guests arrived via a circular drive, then passed through a large oak door and up a wide, stone stairway. In the center of the stone handrails, water streamed down rills, splashing into little pools at every landing. The handrails were embellished with red silk roping, and a tapestry ornamented the ceiling above the stairs. On reaching the first landing, visitors gazed out through arcades to the bewitching courtyard. The home's filigreed windows were arched replicas of those in Granada, and the walls were inset with niches for sculptures, fountains, and plants. One red-tiled alcove, bearing an Arabic inscription, housed a copper and brass brazier; Kate's guests never suspected that the red tiles were New York State sewer tiles, a provenance that she found very amusing.

The furnishings and surfaces at Clones all accommodated Kate's below-average height, and though her palatial home had many rooms, none attracted more comment than her own bedroom. From the chamber, located on the second floor, Kate could climb a narrow, winding staircase to a tower affording sweeping views of the whole estate. Since she lived and worked surrounded by people, this private retreat must have seemed like her personal paradise. Kate's bedroom was also invasion-proof, reached by stairs that could be pulled up from above. In her cloister, she was safe from any siege, like the one that threatened Lytton's Leila in Granada. It is tempting to think that Kate might have designed her boudoir to prevent anyone up there from leaving, but in reality, while her palace had drama, it was never of that particular variety. Kate and her siblings were very Victorian in their personal lives. She once gave her favorite niece, Ellen Gleason Boone, a check for a hundred dollars because, as she explained, "I have it on good authority that you don't 'neck' and I want to reward you for comporting yourself as a lady should." Ellen promptly bought a used sofa, two chairs and an oil heater, fixed up a private space in the family's attic and dubbed it "The Neckery."[27]

When Kate spent several months in Japan in 1917, there were rumors of a romance, but she avoided marriage like "the plague," she said, perhaps because of her fear of gold-diggers. By 1924, Kate told a reporter from the San Francisco Examiner, she had received two hundred letters containing proposals of marriage—many of them "unforgettable," she said—but she always staunchly preserved her independence. Although Kate was often described as "mannish," undeservedly so, there was never any whisper or evidence of a "Boston marriage." She enjoyed the company of men and loved the excitement of flirtations, since "they brighten the field of life and incur no lasting obligations. Keep alive romance," Kate counseled, "that spark that gives life perspective and form."[28] But businesswomen, she firmly believed, should avoid marriage—especially early marriage—since one "can't be faithful to the home and husband and the demands of business at the same time." As her philosophical laundress once advised, "You will do well to marry late. If you get the right man you are glad you waited, and if you get the wrong one, you don't have so long to suffer." Kate elected not to suffer at all; she wanted only one thing—"to demonstrate that a business woman can work as well as a man." To emphasize her self-determination, she had one of her Moorish chairs carved with the Latin motto Possum volo, "I can, if I will."

She entertained often under Clones' red-tiled roofs, and numerous dignitaries passed through its lacey arcades and rooms decorated with tapestries, rugs, and exquisite fabrics collected during her frequent travels.[29] She threw picnics at Clones in the summer for Gleason employees, entertained houseguests from out of town, and hosted large parties for business associates and friends including Andrew Dickson White, the first president of Cornell University, and his second wife, Helen Magill White. Andrew White, who was Cornell's president from 1866 to 1885, also had a distinguished career as a diplomat and politician, serving as a New York State senator, minister to Russia, ambassador to Germany, and president of the American delegation at the Hague Peace Conference. In 1890, after the death of his first wife, he married Helen Magill, the first woman in the United States to earn a Ph.D. degree. Magill, whose father was president of Swarthmore College, earned her doctorate in Greek from Boston College in 1877 and met White ten years later, when she was presenting a paper to the American Social Science Association. After they married in 1890, they maintained homes in Ithaca, N.Y., and Kittery Point, Maine, and on August 2, 1910, Kate invited fellow Cornellians to meet them at a reception in their honor she hosted at Clones.

Kate was hospitable to young neighbors, too. In 1914, Alice Wynd, then ten years old, walked with her brother from her home on East Avenue up to "Miss Kate's" house, where they were sent on an errand. The youngsters were afraid of Kate and never would have gone on their own, but their indomitable neighbor greeted them kindly and showed them her house, which they had always considered a bit mysterious. Kate even led them up the stairs to her bedroom, demonstrating the marvel of her retractable stairs. With her ruddy complexion, short white hair, and twinkling blue eyes, she didn't impress them as pretty or ugly, but she was clearly a "presence."[30]

NOTES

1   William Gleason to Kate Gleason, from the Victoria Hotel in Manchester, England, October 28, 1898.

2   Ibid., November 1, 1898.

3   Ibid., from the Central Station Hotel in Glasgow, Scotland, November 7, 1898.

4   Ibid., from the Savoy Hotel in Berlin, Germany, November 11, 1898.

5   Kate contributed a thousand dollars to the University of Rochester between October 1900 and October 1902, and an additional $250 in 1903.

6   Bennett, "Kate Gleason's Adventures," 172.

7   Ibid.

8   Colvin, *60 Years with Men and Machines*, 73.

9   Bennett, "Kate Gleason's Adventures," 168. Many people made references to Kate's clothes, and in photographs she is elegantly turned out. Nevertheless, the story persists in Rochester that she was often clothed in severe black, and there is a photograph of her dressed in that style in 1886, posed with all of the Gleason employees in front of her father's factory. Black may have been a practical color to wear in a factory in those days, but people who knew Kate in Beaufort, S.C., remembered her as always wearing white. It is likely that she dressed to fit the occasion.

10 Bennett, "Kate Gleason's Adventures," 43.

11 Lacey, Ford, 238.

12 Kate Gleason, letter to the editor, *New York Times*, May 21, 1910, 8.

13 Colvin, *60 Years with Men and Machines*, 73.

14 Bennett, "Kate Gleason's Adventures," 43, 168.

15 Kate Gleason, journal, 148. There were fifty cars on Rochester streets in 1901. (McKelvey, *Rochester, The Quest for Quality: 1890–1925*, 184.)

16 Jean France, "Fabulous Factories," *City Newspaper*, September 24, 1987, 1.

17 All of the Gleasons living today are descended from Andrew and Edith Gleason.

18 Ellen Gleason Boone, in discussion with the author, April 1986.

19 Kate and her father, William, were both listed as contributors in the proceedings of NAWSA's Thirtieth Annual Convention, held in Washington, D.C., in February 1898. In 1909, Kate was featured as a speaker at a meeting of the Women's Political Club (later renamed the Political Equality Club), co-founded by Susan B. Anthony's sister, Mary Anthony. The group vigorously advocated placing women in positions of authority.

20 *New York Times*, January 18, 1900, 5.

21 Both volumes are housed in the RIT Archives Collections of the Wallace Center at the Rochester Institute of Technology.

22   Harper, *Life and Work of Susan B. Anthony*, 1382.

23   William Gleason to family, from Lick House in San Francisco, California, April 17, 1906.

24   William Gleason, from the Imperial Hotel in Portland, Oregon, April19, 1906.

25   Ross, "America's Pioneer Woman Machinist," 29.

26   Kate Gleason to George Eastman, June 13, 1911. Eastman Kodak Company.

27   Boone, discussion.

28   "Woman Sets New Styles for Homes," *San Francisco Examiner*, March 5, 1924, 9.

29   In the late summer of 1906, Kate boarded the Ryndam and traveled to France and Norway to visit customers, returning home in late September aboard the United States. In mid-August of 1909, she traveled to Spain and Morocco aboard the Carpathia and returned to New York City from Naples aboard the Roma.

30   Alice Wood Wynd, in discussion with the author, 1987.

CHAPTER 7

The Breakup

B Y 1905, THE RELATIONSHIP BETWEEN JIM AND KATE, so close and supportive in their early years, deteriorated under the stress of managing the business. Kate took a genuine interest in the people she dealt with and was attentive to their business problems; her natural approach was to offer pricing flexibility, a concept that her brothers strongly opposed. Jim and Andrew believed that a set price, internally driven, was a fair price to all concerned; it was a matter of principle to them. Kate's reasoning was more pragmatic. She never lost sight of the Gleason Works' responsibility to maintain a prospering business for shareholders and employees, but her eye was on future relationships. Her approach was more customer-driven than her brothers, and there were clashes.

Her brothers, especially Andrew, were also annoyed by her flamboy-ance and domineering personality. Andrew was a very able engineer, but since he was not involved in sales and did not travel for the firm he was not as well known within the industry as Kate and Jim were. In fact, Kate's salesmanship abilities were held in low regard by Andrew, who viewed her as a "self-promoter," which indeed she was. At least two of Andrew's seven children have described him as a cold and aloof man, whose temper was leg-endary. It is certain that working with Kate at the family company became intolerable to him and his resentment toward her smoldered.

William, meanwhile, graciously retreated from the firm's leadership, believing that his children needed the freedom to manage the concern without his interference. He kept the title of president and went to the plant daily, but William relinquished responsibility for the company to Kate, Jim, and Andrew. There were complications, however. Although Jim was now in charge as general manager, his siblings and father were stakeholders, too, and they all had to seek consensus on major decisions. The fact that Kate was the eldest and not a bit shy about taking charge did not make things easier, and she surely grated on Jim's nerves. She had assumed the role of matriarch when her mother died, clucking lovingly but assertively over her brothers and sister. Kate was never secretive or devious about her manipulations, confident that her siblings would ultimately see the logic of her advice without chafing at her high-handed, dictatorial style. She trusted Jim to make good decisions; like Kate, he harbored few self-doubts, and he had always been strong enough to hold his own with her, never seeming jealous or cowed. Eleanor, on the other hand, lived most of her life in the shadow of her famous sister, and there is no question but that she chafed under Kate's domination. It was Andrew, however, who most resented Kate's imperiousness; he was last man on the totem pole and, given his proclivity toward anger, resentment, and negativity, his discontent grew as the years passed.

To equalize their roles, William recommended that his elder three children make an agreement, which they put in writing and signed on June 7, 1909:

> We each agree that when we hold between us the majority of the stock in the Gleason Works, we will have the constitution and by-laws revised so that the management will require the unanimous action or approval of all three on any question of importance.
>
> And that we will discontinue the usual arrangement of a President with subordinate officers.
>
> The definition of the amount or importance of transactions requiring all signatures must be unanimous and will be based on the volume of business of the company at the time this revision in the constitution is made.
>
> If one of us leaves before we put this arrangement in force, the two who will be left will share the management of the company.

Two years later, in 1911, he and his three elder children signed another contract. William promised to sell, assign, transfer, and deliver all of his 1,887 shares of capital stock in the Gleason Works to Kate, Jim, and Andrew in equal shares one day before his death. Until that time, the agreement stated, he would keep his shares. William's children, moreover, gave him a hundred thousand dollars at the time the agreement was executed and agreed to pay him fifteen thousand dollars a year, beginning July 1, 1911, in exchange for his dividends.

Bronze tablet of William Gleason in the lobby of Gleason Works, Rochester, N.Y., 1931.

Kate Gleason "Fun House" portrait, turn of the century.

Side view of Kate in her black satin dress with embroidery on the sleeves.

Portrait of Kate in her white hat, early 1900s.

Portrait of Kate, early 1900s.

Andrew C. Gleason, Vice President, Treasurer, and Director of the Gleason Works, early 1900s.

Portrait of Eleanor Gleason, as a librarian at the Rochester Mechanics Institute. Painting by Stanley Gordon prior to World War I.

James E. Gleason, 1938.

Parade in front of the Gleason Works, October 1916.

Left to right: James, Miriam, and Emmet Gleason, early 1950s.

Kate Gleason signature on bank note.

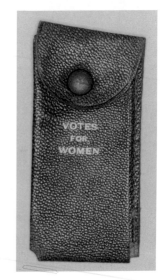

Anna Howard Shaw pouch "Votes For Women," 1918.
Both the note and pouch are courtesy of the archive collection,
Lavery Library, St. John Fisher College, Rochester, N.Y.

A. L. Westgard, Vice-President & Director of Transcontinental Highways of the National Highways Association. He is pictured using one of Kate's campers on his seventeenth motor trip across the United States.

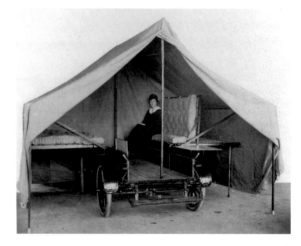

Mrs. Andrew Gleason seated in the camper designed by Kate.

Hi-Speed trailer-car parked in front of Clones.

Trailercar used to transport passengers.

Trailercar used to transport livestock.

Trailercar used to transport mail.

Trailercar used to transport a piano.

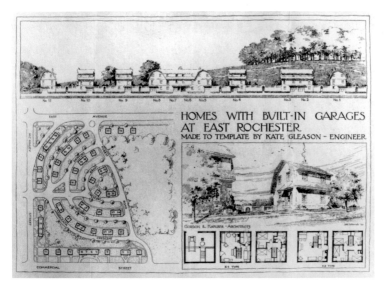

Concrest layout, early 1920s.

All photos of Concrest houses and trailercars are courtesy of the East Rochester History Office, East Rochester, N.Y.

Ad for Marigold Gardens, 1924.

Stanley Walter in front of Genundawah Court, Concrest, 1925-1930.
Photo courtesy of Richard Walter.

Elizabeth Walter at far end of the ballroom of Genundawah Court, Concrest, 1925-1930.
Photo courtesy of Richard Walter.

Despite his best intentions, however, William's concept of having three company bosses with equal authority was fundamentally flawed and ended badly, at least for Kate. She had a high profile in Rochester and the industry, a condition that suited her purposes but galled her siblings. Publicly, she said that publicity proved that women could do business as well as men and opened doors for her when selling machines. By nature, however, Kate basked in the attention, while her siblings were mortified by her affinity for the limelight. Kate's conspicuous home and unorthodox behavior, as well as the distorted stories about her accomplishments, outraged their innate conservatism. Jim, by contrast, did whatever he could to avoid attention; during the Great Depression, for example, he hesitated to buy a new car, as he did every year, inasmuch as so many people had lost everything. He resolved his dilemma by buying a new Cadillac and painting it with a special finish that made it appear old. His older sister, with her taste for publicity and dramatic gestures, was an embarrassment.

In 1911, the Gleasons faced additional changes. William married for the third time, to Margaret Phelan, a distant cousin from his father's side. William and Margaret had traveled to Europe together in 1910, and Kate arranged their marriage, according to family accounts, to put her father out of reach of gold-diggers. William and Margaret traveled widely, spent the winter months in Florida, and enjoyed reading, golfing, playing cards, and seeing friends. Kate loved Margaret and called her "Mother."

That year, the last of the Gleason businesses at Brown's Race finally moved to University Avenue, and Jim was heavily engaged in designing new improvements in gear engineering. His patents and management were instrumental to the success of the business, while Kate continued her sales activities and served as chief financial officer. Andrew, as vice president, mainly occupied himself with production and the design of jobbing work.

Cold, resentful and uncongenial, Andrew was not a happy man. He had few friends and complained often about people he couldn't abide. He was also a tyrannical parent, and Edith tried to shield her children from his anger, hoping to keep him ignorant of their missteps; despite her efforts, however, Andrew threw his second son out of the house three times. His grandchildren, too, remember him with fear and dread; they were never allowed to share meals with him and were always made to feel that they were in his way. Andrew, his daughter Ellen recalled, was "a master at mishandling adolescent boys. He treated them as if they were still eight-year olds" and "was just great at crushing someone's confidence."[1] The reader should know that five of Andrew's seven children were boys.

Her father, she added, was consumed with self-hatred and jealousy and perhaps felt unappreciated and overlooked. Ellen explained that his prejudices were deep-rooted and intractable, especially against Italians and the Catholic Church. John, the eldest of his sons, enraged Andrew by declaring that he had become Catholic and by bringing an Italian college friend home with him one weekend. That was the equivalent of waving a red flag at a raging bull. Andrew detested all Italians because he believed that a few had fraudulently claimed disability in the factory and received compensation. When Edith realized that John's friend was Italian, she quickly rented a hotel room for him, knowing that Andrew would not hesitate to create a scene in front of the young man if he stayed in the house.

By 1913, when the tension between them became unbearable, Andrew threw down the gauntlet and gave Kate an ultimatum, insisting that she was overbearing and difficult to work with. If she didn't leave the business, Andrew said, he would. What triggered the confrontation is still a mystery, but the rift was so poisonous that Andrew forbade his family to utter Kate's name inside his house. From then on, they never did except with contempt, although Edith, to her credit, quietly maintained a lifelong friendship with her sister-in-law.

William must have been crushed when the schism occurred, but he remained neutral and agreed to accept whatever decision his children reached. It was up to Jim to tip the scale, and he sided with Andrew. Kate's choice was to leave the business she had nourished since childhood or tear the company and family apart. "It was heart-breaking," she later acknowledged, "because it meant leaving Father and all the friends I loved." Still, it seemed to her that her wide experience would make it easier for her than for her brother to transition to a different business.[2] She attributed the breakup to the firm's success, which, she claimed, left the founders with little to do, but that is probably too simplistic an explanation for the dramatic break. Andrew's outburst was astounding. Though his complaints were almost petty, he sputtered on about Kate to the end of his life, ingraining in his children the notion that she was more hindrance than help in the business and a fraud to boot. Since the only living Gleasons descend from Andrew, that became Kate's story within the family.

According to Andrew's children Ellen Gleason Boone and Larry Gleason, he leveled three main accusations against his sister. The first was that she had some of her construction bills for Clones sent to her at the Gleason Works—implying that she paid them out of company funds. There is no shred of proof that substantiates or refutes these accusations, but the evidence that does exist suggests Kate was never anything other than scrupulously honest in her dealings.

Andrew's second charge involved a patent dispute between the Packard Motor Car Company and the Gleason Works. The companies argued over which was first to invent a particular method and machine for manufacturing curved tooth bevel gears. The patent office ruled in favor of Gleason, a decision that Packard challenged in the District of Columbia Court of Appeals. Before hearing the final verdict, the Gleason Works and a group of twenty-three automobile and automotive bevel gear manufacturers purchased all of Packard's patent rights relating to bevel gear production. Subsequently, the Court of Appeals ruled in favor of Packard. Andrew claimed that at some point, when the dispute was before the courts, Kate told her friend, the president of the Packard Motor Car Company, that she knew that Packard had had the idea first. Kate cannot answer the charge, but if there is any truth to it, it would have been an uncharacteristically witless and indefensible error on her part.

Andrew's third complaint pertained to the poor state of the firm's financial records when Kate was treasurer. In 1913, the Sixteenth Amendment to the Constitution established the income tax as a permanent fixture in the U.S. tax code. Each Gleason Works department did its own purchasing and billing and kept its own set of books, a system that may have worked well in the early days when the business was small and relatively simple. By 1913, however, that process was hopelessly antiquated. When Andrew assumed Kate's responsibilities after she left the company in 1915, auditors declared the books were so disorganized that they were impossible to audit. This was definitely Kate's area of responsibility, and the auditor's report verifies Andrew's complaint. The patent and bookkeeping issues arose, however, after Kate had left the Gleason Works, so they had nothing whatever to do with her departure.

Once Kate decided to leave the family firm, the Gleasons had to determine how to extricate her from the company with enough money to go into another business. In July 1911, when Kate, James, and Andrew had set up a voting trust, the three siblings owned the majority of stock between them. The insistence on unanimity ensured that if one sibling wanted to sell, he or she would be disposing of an interest that was essentially as good as majority stock. When Kate sought an attorney's counsel about transferring her voting rights to an outsider, however, she was told that it was not legal to make an agreement requiring unanimity and that the voting trust they created was invalid. Kate then proposed a give-or-take exchange with her brothers, for six hundred fifty thousand dollars of seven percent preferred stock, but they considered this deal too expensive for either side. As a counterproposal, William suggested that Kate take a ten thousand dollar income in preferred stock, leaving the balance in common stock. Kate rejected the proposal. "It would have seemed to me exceedingly fair," she protested, "if I were leaving business voluntarily for a life of leisure and could return at any time for association with friendly directors. But as I have always intended to continue in business as long as I live and must now make a new business for myself, I want all my capital available if necessary."[3]

In July 1914, the brothers sent William to Kate with an offer to sell some of their current holdings for six hundred thousand dollars in cash. They proposed that each of the three siblings would sell six hundred thousand dollars worth of stock, leaving their father's holdings intact. Kate considered this proposal fair and contacted Augustus Peabody Loring, a respected Boston attorney, to find buyers for their stock. Kate made clear to Loring that Jim should be kept on to run the business; "If he were released," she explained, "he would start a machine shop of some kind in this city and he would naturally induce some of the men to go with him and they would be the ones the business could not afford to lose. You can be sure that I would not interfere, even if I sold out all my stock later on."[4]

Six months later, on January 19, 1915, William, Kate, Jim, and Andrew signed an agreement stating:

> In consideration of one dollar and other valuable considerations to us paid by Augustus P. Loring, we, the undersigned hereby agree to deliver to him or to his order, on call at any time before May 1st, 1915, twenty one hundred and seventy shares of the Gleason Works of Rochester, New York at a price of five hundred and fifty dollars per share. It is understood that only the ordinary dividends and salaries shall be paid and that said Loring shall make no public offering during the life of this option.

The value of the combined shares was $1,193,500. Within months, however, the family bought back all the options from Loring, after he suggested that the Gleason Works add worm gears to its line. Loring did not understand the difference between bevel gears, manufactured by the Gleason Works, and screw-shaped worm gears, a competing technology that was difficult and expensive to produce and impractical for the automobile industry. Kate quickly traveled to Boston and asked Loring to surrender the options, which he did with regret. Kate, in the end, left the firm without selling her holdings, and the Gleason Works remained intact. By early April, Andrew wrote Jim, she had severed her ties with the company and left her brothers "strictly alone … to sink or swim without her help."[5]

She was moving on. Although her family and business life had collapsed, her professional world was expanding rapidly. In 1913, she had been elected to membership in Verein Deutscher Ingenieur (VDI), the German Engineering Society. She was the first of her gender to join that august group, a distinctive honor for an American. In early March that year, she had sailed aboard the steamship *Prinzess Irene* to visit the Riviera, and in July she was in Cologne, Germany. In January 1914, a bankruptcy court appointed her receiver of the Ingle Machine Company of Rochester, and Kate subsequently assumed management of the firm. Newspapers reported that she was the first American woman to be named a receiver; whether or not that was true, Kate was one of a very few in that position, and a letter to the *New York Times* in 1915 described her as "one of the best mechanical engineers today" who was known "wherever machinery is used in the United States."[6]

The Gleason Works had a financial interest in the Ingle Machine Company. Founded in 1904 by Arthur H. Ingle, the business had been financially overextended and forced into bankruptcy on December 27, 1913, when its two major creditors, Merchants Bank and the Gleason company, called in their loans. When Kate assumed responsibility for Ingle it owed $140,000, and its stock was worthless. Under her careful management, the company paid off its debts, and the bankruptcy was discharged on February 2, 1915. Within two years, it was reported, the firm made a million dollars.

Kate's talents attracted other honors. In 1914, she was elected into the American Society of Mechanical Engineers (ASME) as the first and, for many years, the only woman in that organization. "We have not admitted Miss Gleason as a favor," explained James Hartness, ASME's retiring president. "But she has qualified in all respects as an engineer of the ablest type, who knows not only mechanism but also the rules of business arrangements."[7] Oddly, it is recorded that Kate was elected to ASME for designing a worm gear, something that she never did. Exactly how this inaccuracy came about remains a mystery. There were other criteria, however, and she was lauded as "one of the best mechanical engineers in the country and a manager of finances second to none." She received ASME's encomiums, the press reported, wearing a fashionable evening gown and a pearl necklace.[8] On December 3, 1914 she attended an ASME progressive dinner dance at the Hotel Astor in New York City, the only woman member among the four thousand members in attendance.[9]

Kate was looking ahead. As a tool-company president once told her, "anyone who can make a living out of the machine-tool business could make a fortune out of anything else."[10] She would prove him right.

NOTES

1   Ellen Gleason Boone, in discussion with the author, April 1986.

2   Bennett, "Kate Gleason's Adventures," 173.

3   Kate Gleason to Augustus P. Loring, November 28, 1914.

4   Ibid.

5   Andrew Gleason to James Gleason, Apr. 6, 1915.

6   D.R.C. (Bridgeport, Conn.), letter to the editor, *New York Times*, March 2, 1915, 5:8.

7   "Woman Is Admitted to the Engineers' Society," *Syracuse Herald*, December 2, 1914, 4.

8   "Woman is Engineer," *Ludington (Michigan) Daily News*, January 17, 1915, 10.

9   "Danced and Ate in Relays," *Brooklyn Daily Eagle*, December 4, 1914, 10. At this type of progressive dance, a participant and his or her partner ended at a different table after each dance.

10   Bennett, "Kate Gleason's Adventures," 171.

PART TWO
ANOTHER LIFE
GIVEN

CHAPTER 8

Trailer Cars and Bank Notes

Clones
Pittsford
New York

Thanksgiving Day, 1915

Dear Friend,

This is the Thankfullest Thanksgiving I've had so far. You will be glad to hear it and to know that I am wishing you too, Good Luck.

What makes me so happy is my new work. It is like having another life given, to adventure into such different fields with the advantage of having with me so many of the friends made in the old one.

I hope you can come to see me soon.

Very cordially yours,
Kate Gleason[1]

By the fall of 1915, Kate's horizons had widened well beyond the machine tool industry. Her "Thankfullest Thanksgiving" was as much a celebration of having the means to pursue her dreams as it was of her new life. With war raging on the European continent, machinery orders were pouring in and factories hummed. The calamity engulfing Europe brought prosperity to Gleason shareholders, and Kate's holdings were large.

Her new work focused on three ambitions: promoting the industrial growth of the small community of East Rochester, building low-cost housing for its labor force, and simultaneously developing the manufacture and marketing of trailers pulled by automobiles. Characteristically, she wove these interests into a symbiotic whole with no trace of timidity or second guessing.

Kate had long been attracted to the idea of constructing small, afford-able houses for workers, a goal she had discussed with Augustus Loring in 1914. Now, Kate had the resources to realize her vision in the small com-munity of East Rochester, which was growing rapidly but lacked low-cost housing. In the summer of 1915, she had quietly and relentlessly bought up available properties in East Rochester until the initials "KG" dominated its assessment records for the year. Until the late nineteenth century, East Rochester had consisted of farms bisected by New York Central Railroad tracks. The Vanderbilt Improvement Company, named after the railroad's president, erected car shops for building and repairing railroad cars and gradually transformed the farmland into the village of Despatch.[2] Under the company's stewardship, it became one of the few planned communi-ties outside Washington, D.C., featuring a handsome railroad station, de-signed by renowned Rochester architect Claude Bragdon in 1914; a block of three-story buildings housing a department store, bowling alley, and bar-bershop; and ten buildings erected by the Foster Armstrong Piano Com-pany for manufacturing pianos.

Two of Kate's friends—Edmund Lyon, vice president of the Vander-bilt Improvement Company as well as one of its founders, and Harry Eyer, president of the First National Bank of East Rochester and a Vanderbilt stockholder—had been instrumental in promoting and developing the community. Lyon was a particularly talented and self-effacing individual. An attorney trained at Columbia University, he was also a businessman, in-ventor, and philanthropist. As managing director of the Northeast Electric Company, which eventually became the Delco Division of General Motors, he had invented a mechanism for starting automobile engines that elimi-nated the need for a hand crank. Lyon also devised a locomotive turntable powered by the train engine itself, a breakthrough that was of great im-portance to the railroad industry. In addition, he invented devices to aid the deaf and blind—including one that he specifically designed for Helen Keller—and collaborated on many projects with his good friend Alexan-der Graham Bell, who remembered Lyon as a "big-hearted and big-headed" man who "took my heart."[3] In 1916, Kate honored Lyon's contributions with her own philanthropic gesture. After acquiring seven acres of swampy land in East Rochester, she developed it into a leafy park, which she pre-sented to the community with the requirement that they name it for Ed-mund Lyon. No one was more surprised at the honor than the unpreten-tious Lyon, who would have preferred that the town name the tranquil green after Kate Gleason.

Kate was continuing to acquire property in East Rochester. In April 1916, she and Harry Eyer purchased a thirty-acre tract and a thousand building lots. According to newspaper accounts, Kate had contracted to build three fireproof factories on the thirty-acre parcel and already had occupants for them, including the Erie Mop and Wringer Company, the Merkel Motor Wheel Company, and the E. N. Bridges Photo-Mount Company. She also transferred 315 lots to the village of East Rochester, and the contracting firm of Ransome and Smith transferred eighty-five more. The proceeds from these lots, valued at some three hundred dollars each, provided a fund to promote development of the village.

Kate also set aside a portion of her thirty-acre tract for a plant she intended to build for the Rochester Trailer Car Company. She was intrigued by the challenge of improving hauling transportation and saw potential in the manufacture of trailer-cars pulled by automobiles—an efficient and low-cost way of moving goods and materials that she considered as great an improvement on the wagon as the automobile was to the horse. Her partners in the venture were Charles H. Babcock, president of the Lincoln National Bank of Rochester; Thomas J. Northway, a Rochester automobile dealer; and Emil E. Keller, a vice-president of ASME. Keller had served as vice president and general manager of the Westinghouse Electric Co., and was the engineer responsible for the remarkable feat of illuminating all the grounds and buildings at the 1893 Chicago World's Fair, the world's first major use of alternating current. On February 1, 1916, he had applied for a patent for a trailer steering mechanism, which was granted in August 1919.

Keller agreed to be the trailer car firm's general manager, and in April 1916, he was elected to the Rochester Chamber of Commerce, as Kate had been the month before, both listing their business as the Rochester Trailer Car Company. Kate was the first woman ever elected to the organization, and her foray into the trailer car business, too, was pioneering. In 1914, a Detroit blacksmith and wagon maker named August Fruehauf had built a trailer to pull a friend's boat, and the first advertisements for Fruehauf trailers appeared in April 1916, the same month that the Rochester Trailer Car Company published its first advertisements. Fruehauf and Kate, it appears, were on parallel paths.

Some of the trailers Kate manufactured were used for hauling pianos made at the nearby Piano Works, and others had a box design, with or without sides, for carting cattle or goods. She also saw recreational potential for her haulers, and one of the models she built was a camper, which she personally designed. The camper, her company claimed, was a "practical and luxurious movable hotel on wheels," and some models featured passenger seats.[4]

From the beginning, Kate's ads noted that her trailers were designed and built by the ablest engineers in the automobile industry. In 1918, one of her employees, George Hiller, applied for two patents, one for a draft-bar and the other for a draft-gear for trailers. The patents were granted in 1919 and assigned to Kate, although she had advertised the inventions as early as April 1916. Mounted on regular automobile springs, axles, and wheels and attached to the automobile with a shockproof draw bar, her specially-designed trailers could safely carry up to twelve hundred pounds of merchandise and travel up to speeds of thirty miles per hour. Kate claimed that hers was the only trailer car that tracked perfectly and could be backed in any direction.

In 1916, she was elected vice-president of the new Trailer Manufacturers' Association and in 1917 took the reins as her company's general manager.[5] Under Kate's management, the Rochester Trailer Car Company—which became the Northway Trailer Company in 1917—flourished financially, and Kate was proud of the products she produced. "The business is developing," she told an interviewer in 1918, "and the public is being educated to the use of a trailer car. It is selling," she stated, and her products, the reporter added, "are being sent far and near in carload lots."[6] She was even cultivating markets overseas. On Kate's passport application for a trip to France in the summer of 1919, in fact, she listed her reason for traveling abroad as selling trailer cars.

By 1922, however, her interest in trailers had waned, and in May that year the *Rochester Democrat and Chronicle* carried a "going out of business" advertisement for the firm, offering customers a complete trailer car for the fire-sale price of one hundred twenty-five dollars. The next year, the company was no longer listed in the *Thomas Register of American Manufacturers*, and in 1925 Kate sold the factory building to the Crosman Rifle and Crosman Seed Companies.[7] August Fruehauf, by contrast, continued to focus his efforts on trailers. His name still appears on trailers and commercial vehicles, and he is considered the grandfather of the trailer industry, while Kate's pioneering place in the industry has been long forgotten.

While Kate was busy building new factories and filling them, her brothers carried on with the management of the Gleason Works. Since Kate was safely out of the way, Andrew expected his life at the company to be more serene. He had been contemplating retirement in the future, but in early 1916 he abruptly decided to leave the firm.

On January 21, 1916, from the Hotel Seville in New York City, Andrew wrote to Jim about his plans.

> I told Edith Wednesday noon that I expected to retire from the business within a year but did not anticipate the possibility of doing so within an hour. I realize that it is a very serious move but I do not regret it as I value my peace of mind more than any pecuniary gain I would get out of sticking along under the circumstances.
>
> Kate will get some satisfaction out of it but she will not be any the richer if I can help it. I have often said that you are entitled to much more than one third of this business because you have earned most of it. But with all due modesty I claim to stand next. For the same reason that you deserve most of the credit I have felt that it was up to me to accept things just as they came but lately the situation got to be too much of a wear and tear on my nerves. I am thankful that I am not leaving under the strained relations that Kate did and I will continue to have the same interest in the success of the business as when I had an active part in the management.[8]

Just what "circumstances" disturbed Andrew's peace of mind, precipitating his decision to retire "within an hour," is unclear. As it turned out, however, Andrew did not retire for another eighteen years, so Jim likely persuaded him to stay. At about this time, Jim and Andrew also made a pact that only one son from each brother would be able to work in the family business. Since Jim only had one son, while Andrew had five, Jim clearly got the better of the arrangement; it may have been his price for agreeing to working conditions that would persuade Andrew to remain at the Gleason Works.

Kate, meanwhile, was moving forward, wasting no time on recriminations. After entertaining some of her family at a New Year's dinner at Clones, she wrote to Helen White about scheduling a visit to Ithaca. Kate was an active member of the Women's Advisory Committee of the Federation of Cornell Women's Clubs and was hoping to improve housing there for women students, who were still obliged to live in Sage College under a matron's supervision. In February 1916, Kate attended a Federation meeting in Ithaca to try to find ways and means for providing additional residence halls for women students. The group hoped to adopt a comprehensive plan, and Kate offered a hundred-dollar prize to the member of the senior class in the College of Architecture who came up with the best design for the new facilities.

Some alumnae felt that women students' needs, in general, received short shrift from Cornell's male-dominated board of trustees. To win a stronger voice in university decisions, the Federation's executive committee nominated Kate for the office of alumni trustee, one of four candidates vying for two positions. The group circulated a letter of support for Kate's candidacy to all Cornell alumnae—a move that so alarmed the chairman of another candidate's committee that he proposed banning all circular material about candidates. The three male rivals immediately agreed to abide by the proposal, but the women were less eager, pledging only not to send any circulars out after the first of April.[9] Two days later, the ballots were mailed for the two alumni trusteeships, and results were announced on June 17. Kate trailed the men, but she received 1,782 of 11,120 votes—a reasonable showing considering the far greater number of male alumni. The two winning candidates were both returning trustees, and women continued to be underrepresented on Cornell's board of trustees for many decades.[10]

Kate broke through other barriers in 1914, however. That year, she was the first woman elected to membership in the Rochester Engineering Society, founded in 1897. A fierce internal debate raged over her election, but such controversies didn't trouble Kate. She was never timid about thrusting herself into the middle of a hornet's nest if it paved the way for other women, and she was philosophical about gender discrimination within her industry:

> Has my work been made harder because I am a woman? No, I have no hard knocks to report. Indeed, I think engineering must be different from any other profession in that regard. Engineers are as a class so successful, so progressive that they bear no grudges and feel no jealousies for any newcomers into their ranks, whether man or woman. And besides, I had the advantage of being my father's daughter, and he ranked high. Impossible to estimate the help this may have been to me. Associations count for much in any success and unfortunately they are in great degree a matter of chance. Still, when I recall stories told me by women struggling for place in other professions I insist that engineers are in a class apart.[11]

Despite her optimism, however, there was also considerable anti-woman feeling among ASME members. According to her friend Lillian Gilbreth, who later inquired about joining the society, some ASME members disliked Kate simply because she was a woman.[12] Lillian had completed her Ph.D. at Brown University in 1915 on the psychology of management, after receiving her master's degree at the University of California, Berkeley, where she was the first woman to deliver the commencement address. Her husband, Frank Gilbreth, was a mechanical engineer and occasionally consulted at the Gleason Works and Eastman Kodak Company. Frank and Lillian had married in Oakland, California, in 1904, and they became partners in life and work. On the train the day after their wedding, traveling to Frank's home in New York City, he declared to his new wife, "I want to teach you about concrete and masonry." Lillian, who had studied English at the University of California and expected to be an English teacher, had taken very little math and science; she was hardly prepared to delve into those subjects, especially on her first day of marriage, but she proved a worthy student.[13] Frank soon decided to take Lillian along on consulting trips so he could show her interesting construction work. On one of those trips, Lillian met Kate and, to Kate's amusement, sat in the cab of a small steam engine and learned how to run it. By the time Lillian earned her Ph.D., she was the mother of eight of the twelve children that she would have with Frank. Two of their offspring later wrote a bestseller about their unconventional family, titled *Cheaper by the Dozen*, and Lillian became known as the "Mother of Modern Management." She was another of Kate's "plucky" friends.

Kate's own career was more productive than ever. In late 1916 and early 1917, she began building a country club, which she named Genundawah,[14] on some of the property she had purchased in East Rochester. A forty-room, three-story stucco structure, the clubhouse had the appearance of a rambling country estate, with an imposing fireplace, an indoor swimming pool, a ballroom with a polished cypress floor, and handsome furnishings, many of them imported, including oriental rugs and Chinese hangings. Instead of tearing down existing houses on the property, she integrated them into the club's design, relocating them around a central court. Kate also built three tennis courts and a nine-hole golf course on farm property she purchased to the north and west of the clubhouse. The links-style, fifty-acre course was 243 yards long. Designed by Boston's Donald Ross, it was built in a sandy location where snow melted quickly under the spring sun, allowing an early start to the golfing season. After a construction hiatus during World War I, the complex was completed in 1919, attracting three hundred members, including forty residents; Robert Trent Jones, the father of modern golf course architecture, began his career as caddy-master at Genundawah.

Kate was hitting her stride professionally, but, in 1917, at age fifty-one, she was having health problems, including almost continuous menstrual bleeding from fibroid tumors. Her sister-in-law Edith recommended that she see Dr. Howard A. Kelly, a well-known professor of obstetrics and gynecology at Johns Hopkins Medical School. Sometime that spring, Kate underwent a hysterectomy. Recuperating at Clones, she corresponded with friends, directed her growing enterprises, and entertained many visitors, including her old friend, Archbishop Edward J. Hanna, who had married Andrew and Edith in 1906. Hanna had left Rochester in 1912, when he was appointed auxiliary bishop of San Francisco, and he was subsequently designated archbishop by Pope Benedict XV in 1915. Two years later, in early May 1917, he spent a quiet day with Kate at Clones preparing a baccalaureate address that he would deliver that month at the University of California, Berkeley.

Kate also corresponded with Helen Magill White that spring, attempting to persuade her to join Rochester's Century Club, a venerable women's organization Kate especially enjoyed. She and Helen both loved gardens, and Kate reported that she had "the deeds now for the new 26 acres that is going to be a garden of a Pittsford-Brighton-Penfield moghul." The moghul, needless to say, was Kate herself.[15]

By summer, she was feeling well enough to travel to Seattle and take a steamer to Japan to attend an engineering meeting with other ASME members. It was an eventful trip. Kate was gone at least three months, and there were hints of a romantic involvement and "near seduction" when she was in Asia, which she apparently disclosed in a letter home. That letter has disappeared, but there is another reference to Kate's amorous adventures in a letter she received the following year from a knowing old friend serving with the U.S. Armed Forces in France. "Some one told me recently," he wrote, "that you narrowly escaped matrimony in Japan. There isn't anything wrong with your luck."[16]

As it turned out, Kate's luck in Japan also saved her life. On October 1, when she was in Tokyo, one of the most violent typhoons in history swept Japan. In four hours of devastation, the storm left more than five hundred dead, submerged a hundred fifty homes, and left two hundred thousand homeless in Tokyo Prefecture alone. The surging seas submerged a small island off the coast, drowning all of its three hundred inhabitants.[17] When the force of the storm hit, Kate was in her hotel room, and the violence of the tempest nearly sucked her right through the window.

A month later, she was on her way back to Rochester, and her homecoming on November 9 was a civic event. A large welcoming group, including family members and officials from local organizations—the East Rochester Board of Trade, the East Rochester Branch of the Red Cross, the first and second Liberty Loan Committees, and the East Rochester Community Welfare League—greeted Kate when she arrived. Two days later, the *Rochester Democrat and Chronicle* recorded her impressions of Japan. At a time when the United States discouraged Japanese immigration and the West Coast, in particular, was rife with anti-Japanese nativism, Kate declared that the people of Japan were "clean, cheerful, polite and not quarrelsome,"[18] and should be encouraged to immigrate to the United States from their overcrowded homeland to help ease labor shortages during World War I. She also commented that "some bright manufacturers, especially those of knit goods and other textiles," should start factories in Japan. Kate explained that the Japanese were quick to learn, and that the manufacture of almost any kind of clothing might prove profitable. Given her family's passionate interest in voting rights, she also made a critical observation about Japanese elections, noting that "only the handsomely rich vote. Even nobles who are not wealthy cannot vote, neither may college professors exercise that privilege. Perhaps a few more than a hundred thousand Japanese," she added, "are allowed to use the ballot."

Some of her opinions and remarks would prove inaccurate. Quoting Baron Koudo of the Nippon, Yusen, Kaisha Steamship Company, she predicted that "Japan would never make war on the United States."[19] Years later, before the start of World War II, Kate also asserted that Mussolini was good for Italy. In time, events would contradict her views. In 1917, the year Kate first visited Japan, international relations were in profound turmoil. The Russian revolution and Soviet coup filled Americans with apprehension, and after April 4, when the United States declared war on Germany, citizens from all classes and occupations pitched in to aid the war effort. Kate's sister, Eleanor, resigned from her unpaid position as the first trained librarian of Rochester's Mechanics Institute to join the headquarters staff of the American Library Institute in Washington, D.C., and then moved to France to serve as the American Library Association librarian in Neufchateau.

At fifty-one, Kate's friend Harry Eyer was too old to join the armed forces, but he abruptly moved overseas with the YMCA, leaving the First National Bank of East Rochester without a president. On August 16, 1918, the bank's directors met and unanimously elected Kate to that position, acting, in all likelihood, on Eyer's recommendation. At the time, Kate was believed to be the first woman ever elected president of a national bank, but it is now clear that a number of women had held similar positions earlier. Still, Kate was probably the first woman to head a national bank without previous family ties to the institution.[20] Following her election, she received a letter from an old friend who was a captain in the American Air Service in France: "Today's Paris edition of the *New York Herald* had a cable item that you were the President of a bank. Hooray for the Irish," he cheered, adding, "I expect to hear that you have been elected Senator from New York almost any time now."[21]

Kate may not have been running for office, but she did have her eye on a government appointment. Two weeks after she was named bank president, Andrew D. White wrote to Secretary of State Robert Lansing:

> My Dear Mr. Secretary:
> My friend, Miss Kate Gleason, one of our earlier Cornell graduates, writes me that she is about to revisit Japan, and that she has applied to you for a position on the economic and industrial commission to Russia. I take the liberty of recommending her to you most highly. Her rather remarkable career would, I am sure, greatly interest you, if it be not already known to you.[22]

"I may add," he concluded, "that Miss Gleason has an attractive and winning personality which should serve her well with those with whom she is brought into relation." The secretary answered graciously but declined to appoint Kate to the commission.[23]

Still, she had considerable authority. In those days, bank presidents and cashiers signed national bank notes. Perhaps the most astonishing aspect of Kate's new position as bank president was that she could inscribe her signature on U.S. currency, but she could not vote.[24] In recognition of her historic, if not unprecedented, position, she gave the first twenty-dollar bill she ever signed to her friend Dr. Anna Howard Shaw, who had headed the National American Woman Suffrage Association from 1904 to 1915. Shaw kept the bank note in a worn leather case stamped "Votes for Women" and carried it with her wherever she went as a good luck piece.[25] Following Dr. Shaw's death in 1919, a fund of $100,000 was established to endow a chair of politics at Bryn Mawr College in her name. Susan B. Anthony's niece, Lucy Anthony, a life-long friend of Dr. Shaw, presented the twenty-dollar bill to M. Carey Thomas, President of Bryn Mawr, as a donation to the fund. "Kate Gleason was not only a co-worker of Doctor Shaw's, but was a friend of many years' standing of Miss Susan B. Anthony, the pioneer suffragist."[26]

Leading a bank was a radical departure from trailer car manufacturing and the development of East Rochester housing and industry, but Kate considered it her patriotic duty, part of her war work, and she rarely passed up opportunities to strike a blow for her gender. Managing the bank, however, proved more challenging than she may have expected, especially considering that she had no previous experience as a director or stockholder. Before he resigned from the bank, Harry Eyer had left Rochester for several months for health reasons. In his absence, the cashier ran the bank, which was apparently paying dividends in excess of profits; less than six weeks after Kate's election, the cashier suffered a nervous breakdown. The bank's examiner filed reports critical of its large number of outstanding loans and vacuum of leadership.

Kate promptly replaced the cashier, and by February 1919, the examiner reported that the "bank has made some progress in getting settlements on past due paper."[27] The November examination that year noted that more doubtful loans had been paid or written off, and the bank was gradually returning to financial health. Kate, however, may have had too many things on her plate to give bank matters her full attention, including running a trailer car business, building factories and a country club, planning a housing development, chairing the Rochester Chamber of Commerce Flower City Beautification Committee, organizing the Rochester Rose Society, and traveling to the Far East and Europe. The examiner noted that she seldom attended board meetings and that the bank's directors had not met in July or August. In January 1920, she stepped down as president, although she remained a director. Kate's talents, she freely acknowledged, were more entrepreneurial than managerial. "The greatest fun I have in life," she explained in 1919, "is building-up, trying to create." As she later observed, "I was no great shakes as a bank president. The fact that the bank was more prosperous when I resigned than when I was made president was due mainly to circumstances."[28] It was a fair assessment.

NOTES

1 Kate Gleason to friends, November 25, 1915.
2 The name of the village changed from Despatch to East Rochester on October 18, 1906.
3 Remington, *Vibrant Silence*, 320.
4 "Rochester's Trailer Industry," *Rochester Motorist* (Automobile Club of Rochester), July 1918, 15.
5 "Alumni Notes," *Cornell Alumni News* 19, no. 15 (January 18, 1917): 177.
6 *Rochester Herald*, August 18, 1918, 10.
7 The Crosman Seed Company still occupied the building in 2009.

8   Andrew Gleason to James Gleason, from Hotel Seville in New York City, January 21, 1916.

9   "The Alumni Trusteeships: Proposal to Refrain from Circularization Not Accepted by All Candidates," *Cornell Alumni News* 18, no. 26 (March 30, 1916): 307.

10   Conable, *Women at Cornell,* 130.

11   Chappell, "Kate Gleason's Careers," 10, 20.

12   Graham, *Managing on Her Own,* 101.

13   Yost, *Frank and Lillian Gilbreth,* 114.

14   "...This she named Genundewah [sic], and only Kate and the Indians know why ..." from an article by Curt Gerling for the *Brighton (New York) Pittsford Post,* October 21, 1991.

15   Kate Gleason to Helen Magill White, n.d.

16   Captain Charles A. Harmon to Kate Gleason, August 21, 1918.

17   "2,174 Perished in Japan's Typhoon," *Trenton Evening Times,* October 8, 1917; *London Times,* October 8, 1917.

18   "Favors Coming of Japanese to United States," *Rochester Democrat and Chronicle,* November 11, 1917, 28.

19   Ibid.

20   Kabelac, "Kate Gleason, National Bank President," 67.

21   Harmon to Kate Gleason, August 21, 1918.

22   Dr. Andrew White to Robert Lansing, August 29, 1918.

23   Andrew White was in failing health and died before the year was out.

24   The Nineteenth Amendment to the U.S. Constitution was ratified in 1920.

25   Bill and case, Lavery Library, St. John Fisher College, Rochester, New York.

26   "Proposed Memorial for Anna Howard Shaw," *Chester (Pennsylvania) Times,* December 2, 1919, 15.

27   Supplemental Examiner's Report of First National Bank, East Rochester, New York, Remarks, February 13, 1919, National Archives, Washington, D.C.

28   Bennett, "Kate Gleason's Adventures," 174.

CHAPTER 9

Building Up

K ATE'S PRODUCTIVE ACTIVITY BETWEEN 1915 AND 1919 would
have overloaded a less ambitious and energetic person, and through
it all, she was laying the groundwork for her boldest project yet. She had
always enjoyed building more than anything else, and since 1914, she had
set her sights on profitably mass-producing affordable housing.

Kate had gained valuable experience building Clones and during her
brief banking career, when she finished some houses for a builder who
had defaulted on a loan. She had also acquired a portfolio of properties,
including a parcel on a slope across the street from the factories that she
built in East Rochester. It was here that she chose to build her low-cost
housing. Factories needed workers, and employees needed affordable and
convenient places to live. After World War I, there was an even greater need
for low-cost housing across the country. Homebuilding had come to a halt
during the war, when money and manpower were diverted to military uses.
Rental units were costly and hard to find, and new homes were virtually
unavailable. In 1919, when the war ended, a million marriages were performed in the United States, but builders constructed only seventy thousand new houses.[1]

New families needed places to live, and Kate was ready to meet the
challenge in East Rochester with an innovative development she began
building in 1919, which she called "Concrest." With the understanding of
a seasoned marketing executive, Kate identified her potential buyers as low-
er-income apartment renters who had been priced out of the housing market. She knew what her customers needed and could afford and planned to
accommodate them, while making her dwellings profitable to build.

For inspiration, Kate visited small tract housing sites in the United

States and Europe, studying them for styles and methods that she could incorporate in her own buildings. On one such trip in the winter of 1920–21, Kate, still intrepid at fifty-five, traveled from Paris to London in an enclosed, eight-passenger Handley-Page airplane so that she could inspect the types of houses that were being built in Plymouth by the British government; her daring flight across the English Channel was on one of the earliest regular commercial international air service routes anywhere in the world, inaugurated in August 1919, little more than a year earlier.

Kate hired Edwin Gordon, who had designed Clones, and his Rochester architectural firm, Gordon and Kaelber, to draw up plans for her new project. She designed Concrest to resemble a French village, like dozens she had passed through in her frequent travels. Its fifty-seven cast-in-place concrete homes were placed at appealing angles to each other on roads that curved attractively around a hillside. She included duplexes and townhouses among the single-family homes, all built in what she called Dutch Colonial style, since she believed it was more economical and pleasing to the eye to build closely-sited houses with a uniform architectural aesthetic. In this approach, Kate was influenced by the attractive Forest Hills low-income housing development in New York City, which was built in 1911 and modeled on English garden communities.

Kate decided to use concrete for construction instead of traditional wood or brick for three reasons: there was a workforce in East Rochester that was well-trained in concrete construction; the homes would be fireproof; and she owned a remarkably fine sand pit within a mile of Concrest. Her workmen had learned the intricacies of building with concrete when they worked with engineer Ernest Ransome, who had developed a system of reinforced concrete construction, on the American Piano Company factory at the turn of the century. With Kate's sand bank, she was able to charge sand to the job at $1.50 a yard and make money on it at both ends. Finally, her experiences with the terrible fires she had witnessed taught her the value of using non-combustible building materials.

Concrest was not the first concrete home development in the United States, nor was it the first example of tract housing, but it was the first tract of concrete homes in America constructed by a woman, and Kate used innovative methods. She was influenced by the pioneering use of concrete construction by Thomas Edison, a Gleason family acquaintance. In 1906, he had introduced his own patented plans for concrete homes, declaring that they would revolutionize American life. Edison built a sprawling cement plant near the town of Stewartsville, New Jersey, and experimented with mass-producing inexpensive, fireproof, cast-concrete homes. After improving the molds and casting process, he erected his first concrete house in South Orange, New Jersey, in 1911. In 1917, he began mass-production of a forty-house residential development, with homes priced at $1,200, two-thirds less than the average house. Edison's homes, however, failed to appeal to buyers.

Unlike Edison's concrete homes, Kate's were designed with aesthetics in mind. Their exteriors were stuccoed and attractively tinted, and though Kate repeated several standard plans in her Concrest houses, she situated the homes at different angles to minimize the appearance of uniformity. The 20' x 20' houses, on 50' lots, were small compared to other homes of the day, but they were distinctly larger and more complete than an apartment. For the sum of four thousand dollars—on easy terms of four hundred down and forty dollars a month, including taxes and interest—Kate provided a five- or six-room, two-story home with a fireplace and an attached laundry and garage, an appealing alternative for buyers living in four-room, sixty-five-dollar-a-month apartments.[2] Although she made her kitchens quite small, citing 2' x 3' Pullman diners that managed to serve hundreds of train passengers, Kate furnished each with a porcelain-lined gas range, cabinets, a kitchen sink with a special mixing faucet, ice box, laundry tray with aluminum cover, electric ventilator, cook book, ironing board, and floor mats. She declared that if women wanted to go to the movies, "all they have to do is to shove the dishes in the laundry tray, put down the cover and forget them, and maybe if you soak them, in the morning they will be clean any how."[3] Kate completely furnished a model home to show prospective buyers how they could do it, and it was reported, no doubt with amusement, that she furnished each house with a powder puff and mirror to please the ladies. While true, it was men she really sought to please by designing each house with a warm garage, equipped with doors and hangers that cost as much for one home as mirrors and powder puffs did for forty.

Advertisements for her houses were quintessentially Kate:

Hang up your hat in a home of your own and enjoy the thrill that comes when you look over the happy homestead and proudly exclaim, 'It's mine! All mine!' Say a sad farewell to your landlord, who never, never, forgets you, and apply 40 a month on a home of your own—a home with a deed, title, porch light, garage, fine view, fireplace, electricity, green grass, French windows and everything. Pick four-leaf clovers in your own lawn while the children romp in the sand piles or run in the nearby grove of trees. Play

'Home, Sweet Home' on your own front porch this summer while you enjoy a ten-mile view of the surrounding country and the cooling breezes are wafted your way.[4]

She called her homes "apartments with gardens" and encouraged residents to beautify their lots by sponsoring an annual competition for the loveliest plots, as judged by the presidents of the Rochester Garden Club and the Federated Garden Clubs of New York State. Kate also built a park next to Concrest, lined her streets with maple and cedar trees, and gave them names evoking romantic tranquility: Woodneath Crescent, Pomander Walk, and Drumore Crescent.

Kate could sell her houses profitably at a low price because many of her methods saved time and money. Her experiences with assembly lines and mass-production in gear manufacturing, as well as her long-standing friendship with time-and-motion experts Frank and Lillian Gilbreth, poised her to apply these techniques in homebuilding. One of her inspirations, she explained, "came from a visit I made [to] the Cadillac factory a few years ago, when [its president] Mr. Leland showed me the assembly of the 8-cylinder engine. All this work was done by one man, who was furnished with a cabinet on wheels, which contained every part he needed and only as many parts as he needed ... We try here to follow Mr. Leland's methods as closely as possible, by having the stock on the job ahead of time, as needed."[5]

At the Concrest work site, Kate fitted a trailer car with plumbing and electrical outfits, cutting tools, vices, and threading tools and moved it from house to house with a Ford chassis. She had four portable sheds, a portable office, a tool house, and a saw shed that she arranged to form a hollow square. In the middle, she stored her lumber and other materials to keep them safe. She even minimized the waste associated with the loss of small parts by keeping a locked box in which workers placed any wedges and bolts that had been stripped from a form.

Kate disliked the steel scaffolding in use at that time in the United States because when workers dismantled them they invariably let them "go smash," as she put it, and they never seemed to fit together quite the same way again. Instead, she followed the French example of using as many pairs of extension ladders as needed across the front of a house, ladders which were durable, inexpensive, and easy to move.

Kate found much of the machinery on the market too heavy, awkward, and time-consuming for her purposes, since it was geared for projects larger than the small houses she was building. So she put George Hiller to work designing a concrete lifter that would not take five men a day and a half to move, like the one currently in use. The resourceful Hiller, who lived conveniently in a corner house across the side street from Concrest, subsequently patented the telescoping cement elevator as he had the draft-bar and draft-gear for Kate's trailer cars. His cement tower was built of angle iron that telescoped to forty-six feet and could be lowered to nineteen feet. Raising and lowering it took twenty minutes, and when lowered, it could be moved the fifty feet from house to house in just two hours. With this machine, Kate's workers could pour a whole house in fifteen hours, including first and second story walls, stairs, stairwell, a complete chimney, floors, columns, and beams.

To unload cement from a railroad car and transport it half a mile to her storage shed, Kate used three of her hi-speed trailer cars and a Ford chassis. While two men loaded cement into an empty trailer car, a driver took a loaded trailer car to the shed, where two men unloaded the material. Unhitching the Ford from the loaded car, the driver then took the third empty trailer car back to the railroad switch, unhitched it for the men there to fill, and then drove the first (now loaded) trailer car to the shed. Using this method, it took three hours and five men to unload an entire railroad car.

Kate also designed houses that were efficient to build. She eliminated cellars from her homes, which reduced plumbing and sewage pipe costs and eliminated the need for an extra stairway. She built roofs without expensive valleys and was a proponent of one-line plumbing, not yet generally in use; having a bathroom and kitchen side by side or one above another cut plumbing costs, as did having the sink, toilet, and tub head in a single line.

By reducing her ceiling heights from the customary nine feet to seven and a half, Kate adjusted them to the scale of her houses and reduced the number of stair treads between floors from seventeen to eleven. All her houses had a ten-foot width, a span that reduced the need for expensive reinforcement and roofing material. The site's hillside setting also allowed Kate's landscape engineer, Frederick (Fritz) Trautmann, to put two stories on the front of each house and three stories in back, avoiding a significant investment in grading.

Careful buying in quantity was another cost-saving strategy. Kate determined that twenty-five houses was about the right number for each job—the inventory moved faster, reducing money tied up in materials; the duplication of tasks created time efficiencies; and the cost per house declined with each home built until the average cost of production was about 45 percent of the cost of the first house in the job.

Good personnel management was also vital. To keep her men steadily employed, Kate had bad-weather jobs planned ahead for them to do, and she eliminated as much handwork as possible because it was expensive and varied from worker to worker. She also hired contracts for some of the work, such as putting up forms, because she believed that when men felt they were in business for themselves, they did their best. As always, Kate was a demanding but considerate taskmaster. According to one observer, she was "on the job more than any man on the payroll; she knows every workman by his first name, and uses it, and every detail of every part of the work, and personally looks after it."[6] On very hot days, or to show her appreciation for necessary overtime, she would sometimes serve workers cool drinks or ice cream, and on dismal, cold days, she occasionally served the men coffee and doughnuts "without any idea," she tartly insisted, "of being benevolent."

With all these economies in place, unknown and unseen by the would-be buyer, it fell to Kate, a master saleswoman, to add noticeable but inexpensive features to the homes that gave them an appearance of richness: ornamental tile around the fireplace, a clean-out for ashes and a damper, a built-in bookcase, and a brass-trimmed wood box on wheels, as well as window shades, banisters, screens, and even a laundry basket, all designed to appeal to fledgling homeowners.

Always mindful of the value of publicity, Kate wrote and spoke extensively about the project. Articles appeared in journals, magazines, and newspapers. The Petersburg, Virginia, *Evening Progress* reported on August 20, 1920, for example, that

> A woman is leading the way in Rochester's fight for adequate housing—she's an engineer, a master machinist, society woman, banker, and a leader in Rochester's Chamber of Commerce, too. She is Miss Kate Gleason. All her life Miss Gleason has been a trail-blazer—she has led and Rochester has followed. She has enjoyed breaking down traditions and entering fields of endeavor hitherto barred to women.[7]

Kate was invited to give an informal talk about Concrest to the members of the American Concrete Institute (ACI) at their eighteenth annual meeting in Cleveland, Ohio, in February 1922. She nearly missed the event, since her ship was fogbound en route from France and Italy. As soon as she landed in the United States, she wired the institute's executive that she would be on the next train to Cleveland and would speak as promised. Kate asked him to have a hairdresser ready for her upon her arrival, and that same evening, she swept into the hall wearing a Paris gown and hat "for moral support," she said, "in undertaking to tell an audience of men how to build houses." Those in the audience who had seen Kate personally directing the work at the muddy Concrest building site remembered long afterward the no-nonsense and stylishly feminine side of this "charming, resourceful and unusually interesting woman."[8] By June of that year, she had been elected to membership in the ACI. She was the first and, even at the time of her death, the only woman member of the organization. As she told an interviewer, she had been "developing a talent that almost amounted to genius for putting myself in places where other women are not likely to come."[9]

In 1923, Kate began building an extension of Concrest, called Marigold Gardens, consisting of spacious homes set on larger lots. She called the new homes "Ellen McDermott Cottages," after her mother, and advertisements described them as designed by Kate Gleason. These homes, situated in a flat area across the street from Concrest, are attractive and charmingly angled.

In all, Kate built more than a hundred houses in East Rochester. Needing help managing all the properties, she published a help wanted ad for an office manager and maintenance man. A couple, Stanley and Elizabeth Walter, applied for the jobs, and Kate must have hired them on instinct, since they lacked any experience in those positions. The couple had been chicken farmers in Elba, New York, and had also had colorful careers in vaudeville. Stanley Walter and a partner had a novelty act, called the "Whitman Brothers," on the highly regarded Orpheum circuit. They were best known for staging a swamp scene in which Walter played a frog and his partner an alligator. Walter was a contortionist, so the frog amazed audiences by assuming a variety of startling positions.

His wife, Elizabeth, was also an entertainer. She had been known by the stage name "Labelle Stone" and performed extensively in Europe. She was a contortionist, too; in her act, she climbed inside a ball, then rolled herself up a spiral structure by manipulating her body inside the sphere. At the top of the slope, Elizabeth would briefly open the ball, releasing two white doves; then, still tucked inside, she would roll the ball back down the spiral in a controlled descent, the most challenging part of her act. A photo of Stanley Walter outside the Genundawah Court ballroom, with one leg planted on the ground and the other at a 90-degree angle up a post, may explain why Kate hired the couple; they were flexible and, surely, entertaining. Stanley's ability to maneuver in tight spaces added to his skills as a handyman, and he maintained Concrest homes for the rest of Kate's life and for years after.

The Walters soon had a son, Richard, whom they adopted when he was eighteen months old, thanks in part to Kate's generosity and fondness for children. She paid the couple an extra $25 a month to cover the cost of supporting their adopted child and provided living quarters as part of their salary. Richard Walter remembered riding a tricycle as a child that his mother told him was a gift from "Miss Gleason."[10]

In the early 1920s, however, Kate's own financial situation was a bit precarious. She had so much money tied up in real estate that she had to borrow as much as six hundred thousand dollars to continue her building project. Although local banks had agreed to lend her the money, they withdrew their offers when a sharp recession slowed the economy. Desperately searching for some way to continue her homebuilding, Kate turned to a successful young businesswoman in New York City whom she had once assisted. Help finally arrived in 1924 through the young woman's banker, Walter Stabler, who was comptroller of the Metropolitan Life Insurance Company. Cornell University also approved a loan on eight of Kate's houses, but it did not sweeten her relationship with her brothers when she put her Gleason stock up as collateral for the loans.

Kate was a risk-taker, and she came closer to the brink of disaster in those years than she ever did before or after. Some thought Concrest would bankrupt her, and Kate was worried that they might be right. But her houses were good, and Kate knew it; in the end, she was able to pay off all of the loans in just three years. As she explained to Helen Magill White, once her affairs finally straightened out, she would be free to build "six room and garage houses" the rest of her days; even in her early sixties, Kate predicted with unshaken confidence, "the best of my life," and even greater adventures, were still ahead.[11]

William Gleason died on May 24, 1922, at the age of eighty-six, after a two-month illness. His funeral was officiated by President Rush Rhees of the University of Rochester, and the twelve men who were employed longest at the Gleason Works were his honorary pallbearers. From the peat bogs of Ireland, he had risen to be a captain of industry. A decade after his death, a bronze tablet was placed in the lobby of the Gleason Works, commemorating William as "a master craftsman, endowed with indomitable will, a spirit of independence, and broad vision, who created a new type of machine tool and founded this business on ideals of service and fair dealing."[12] It was one of his outstanding characteristics, Eleanor said at its unveiling, "that he never allowed the growing conservatism of age to stand in the way of new ideas, new ways of doing things. That quality, together with his driving energy and his absolute fairness, are the things we remember best about him. We remember, too, some blustery days, for he was rather quick to fly off the handle, but his sense of justice never allowed him to be influenced by these passing gusts."[13]

Nowhere was William's spirit of independence more evident than in his encouragement of his daughters and other women. He urged them to take their rightful places in what was then a man's world. He had supported women's suffrage when it was not popular or common for a man to do so, and Susan B. Anthony had called him "noble."

Months after his death, Kate made a gesture to begin healing the rift with her brothers. That fall, aboard the *Rochambeau,* en route to France, she wrote to Jim, "I have felt so bitter to you both, particularly to you, that it did not seem credible to me that I could ever forgive you. But the incredible has happened and since Father's death I even feel kindly towards you both. It seems his influence is stronger with me now than when he lived."[14]

She refused, however, to have anything to do with the Gleason Works or attend any of its shareholder meetings. "I can't get over my old love of the business," she explained to her brother; "it hurts me to be out of it. If I go there, I will be like the mother-in-law whose only child has married. It is better to be so absorbed in something else that I have a chance to forget it."[15]

So Kate immersed herself in travels and "adventures in building" beyond East Rochester. On April 23, 1923, she wrote Helen Magill White: "My first contract will be in San Francisco. From there I will go on to the Vale of Cashmere for February. Then to England for April where I am to represent the Am. Soc. Mec. Engineers at an English convention of Engineers. And I will be back to America in the summer for more building. Doesn't it sound pleasantly adventurous."[16]

Kate's life and her sphere of interests were fast expanding. The following month, in May, she wrote George Eastman, founder of the Kodak Company in Rochester, declining his invitation to an entertainment because of her ASME commitments in Montreal. "I am particularly sorry to miss seeing you," she added, "as I will probably never have another opportunity. I hope to leave Rochester in July to go on with my work of house building in places where I have contracts and it is unlikely that I will be here again except for a day now and then."[17]

Kate admired Eastman. He had started his career as a junior clerk at the Rochester Savings Bank before launching his own company, inventing the name Kodak and trademarking it in 1888. In 1919, Eastman had donated a third of his shares in the company to his employees, and Kate told him in her letter how much she respected his "action in giving the ordinary stock-holders such a big share of the rewards of your business as soon as there were any rewards." Her letter praised Eastman's "business ideals of courage and fairness." She also mentioned a trial he was involved with, telling him that her heart ached for him and adding, not quite diplomatically, that "you looked some days as though you had been buried about two weeks and dug up again ... I have myself," she remarked, "come through a terribly tough fight in '20 and '21. If my father had not been here to help me, I was a goner. I am glad now I had the troubles otherwise I might have stuck around here and missed the gay life of adventure and congenial work I have before me."[18]

Kate had planned to travel to San Francisco in the summer of 1923, but she postponed the trip for a year because of her health. Her doctor ordered her to take a vacation, so she decided to go to France for a few months, where she looked forward to visiting some friends and learning "something about design that will help me in making my houses here more beautiful."[19] In France, she was planning to take many long walks as part of her life-long battle with the bulge. She was also, she told Helen Magill White, "hoping to find the apartment Anatole France describes in *The Crime of Sylvestre Bonnard*, on the Quais Malaquais, overlooking the Tuileries Garden and the Seine and the Louvre with the turrets of [Sainte] Chapell and the towers of Notre Dame."[20]

Kate sailed for Europe aboard the White Star liner *S.S. Pittsburg* and found congenial fellow passengers, including a U.S. navy captain, a Kentucky judge, a Philadelphia banker, a Providence architect, and the advertising manager of the Curtis Publishing Company, who had conveniently brought along new issues of her favorite magazine, the *Saturday Evening Post*. Kate had been enjoying articles by Herbert Henry Asquith on the causes of World War I, even though she said she "never expected to admire any of Asquith's work. He seemed needlessly obstructive to Woman Suffrage in England," she explained, "and his wife Margot impresses me as quite ill-mannered so I supposed he would be like her."[21]

Arriving in France on August 1, Kate traveled to Paris and renewed her friendship with an old acquaintance, Henry Selden Bacon of Rochester. A major in the aviation section of the U. S. Army Signal Corps, Bacon had served in Paris and London during World War I and, as chief legal advisor to the air service, had attended the historic Versailles Peace Conference. Following the war, Bacon had served as Paris counsel for the Standard Oil Company of California and legal advisor on American matters for French

Prime Minister Georges Clemenceau.[22] An author of numerous articles on Franco-American relations, Bacon was also an officer of the French Legion of Honor. Kate saw a great deal of him in Paris and wrote to Helen Magill White that he was

> ideally happy and intends to live out the rest of his life in Paris. He is prac-
> ticing law there and evidently has lots of hard work to keep him busy. He
> has a charming apartment at the top of an old house overlooking the Seine
> and Notre Dame. He has a good cook and beautiful pictures and lots of
> books and a comfortable little motor car. I wish you could make a trip to
> Paris with me next September when I will probably be going again. It's al-
> most an American city now but we can't quite spoil the French charm.[23]

Her Paris sojourn was interrupted in February, however, when she received a letter from her friend James I. McLaughlin, a transplanted Rochesterian who was working as a contractor in northern California. On September 17, 1923, a wildfire had raged through the Berkeley hills, destroying nearly six hundred homes and scorching a hundred thirty acres. One in ten Berkeley students were homeless, and irreplaceable academic collections and papers were reduced to ashes. In the aftermath of the inferno, local builders were using stucco instead of highly flammable materials like redwood, and McLaughlin asked Kate to advise him on stucco construction methods. It was time, she decided, to attend to the building projects she had envisioned in California, and she promptly set out for the West Coast.

"Two weeks ago I was in Paris," she told a reporter who interviewed her in San Francisco. "Now I am back to work," she said, but "I did get some lovely hats before I left."[24] Kate's influence on Berkeley's rebuilding is unclear, but her impact was certainly felt across the bay in Sausalito. In April 1924, she bought a wooded property in the scenic town, on a hillside above Richardson Bay, and established a residence on a triangular lot at the corner of North and Atwood. That same month, Kate advertised that her Rochester home, Clones, was available for lease.

Sausalito, in the 1920s, was attracting hundreds of day-trippers from San Francisco, the East Bay, and points beyond. Hiking was the rage, and nearby Mount Tamalpais drew hikers and tourists with its views and trails. From Mill Valley, at the base of the mountain, visitors could ride to the summit by rail or hike up its many trails. The Tamalpais Tavern awaited visitors at the top with food, drink, and overnight accommodations. A railroad spur from Mill Valley to Muir Woods also enabled tourists to visit the redwood grove by gravity car, without benefit of a locomotive, with only a brakeman manning the ride at a thrilling twelve miles per hour.

Hikers were not the only adventurers attracted to Sausalito in the twenties; bootleggers found it an inviting haven for illicit activities. Throughout the Prohibition years, from 1920 to 1933, "rum-runners" from Vancouver, obscured by heavy fog, would slip into the harbor and hastily off-load their cargo; notorious gangster Baby Face Nelson lived in the town for a time in the early thirties.

Given the town's picturesque setting and popularity, Kate spied a business opportunity and systematically began buying property in "Old Sausalito," an early residential section separated from the commercial district by a steep bluff. She remained in the Bay Area long enough to know what kind of property she wanted, and she hired a local realtor, W. Robert Miller, to act on her behalf.

NOTES

1   National Housing Association, *Housing Problems,* 320.
2   The new homeowner could count half of the forty dollars per month as a capital investment.
3   American Concrete Institute, *Proceedings,* 126.
4   Kate Gleason, "Hang Up," newspaper fragment ('1921' written in pen).
5   Kate Gleason, "How a Woman," 11.
6   Ibid.
7   Edwin D. Rider, "Woman Shows Way to Solve Housing," *Evening Progress* (Petersburg, Virginia), August 20, 1920.
8   Obituary of Kate Gleason, *Journal of the American Concrete Institute,* February 1933, 6.
9   Chappell, "Kate Gleason's Careers," 38.
10   Richard Walter, in discussion with the author, February 6, 2004.
11   Kate Gleason to Helen Magill White, December 28, 1921.
12   The tablet still hangs in the same location today.
13   Eleanor Gleason, speech, January 29, 1931.
14   Kate Gleason to James Gleason, from aboard the Rochambeau, October 19, 1922.
15   Ibid.
16   Kate Gleason to White, April 23, 1923.
17   Kate Gleason to George Eastman, May 23, 1923.
18   Ibid. In 1920, Kate sold eight hundred shares of preferred stock in the Gleason Works to her father, William.

Libby Sanders, (L) and Carol Cross, (R) in a newspaper article "Two College Girls to Doff Overalls After Season of Regular He-work", 1924.

Postcard of the Gold Eagle Tavern, Beaufort, S.C.,1930.

Tony Agostinelli with Kate Gleason's Ford station wagon at Lady's Island, Beaufort, S.C., 1930. Photo courtesy of Janet Agostinelli Jimenez.

Colony Gardens pamphlet, 1930s.

Interior view of Colony Gardens. "Your large and comfortable living room", Beaufort, S.C., 1930s.

Colony Gardens, "Terrace, badminton court and ninety-foot saltwater pool", Beaufort, S.C., 1930s.

The Gleason House is one of the many unique dwellings in Marin County and perhaps one of the very few that parks the car in the attic. Sausalito, C.A., 1933. © Bettmann/CORBIS

Kate Gleason's house with donjon behind, Septmonts, France, mid 1920s.

Kate's Square Tower, Septmonts, 1926.

The Donjon, Septmonts, late 1920s.

The new bell arrives, Septmonts, 1928.

Interior view of the library Kate Gleason built for Septmonts, 1928.

Kate Gleason (in white) with visitors, Septmonts, mid 1920s.

Kate's house in Sept-monts when she pur-chased it in 1926.

Kate's house after remod-eling was completed in 1928.

Kate's buildings in Septmonts, 1926.

View of the gate entrance to Kate's house in Septmonts, 1995.

Jim Gleason (husband of Janis) atop Kate's square tower with donjon in the background, Septmonts, 1995.

Janis Gleason (L) pictured with Madeline Damas and Marc-Henry Debard (R) in front of buildings once owned by Kate Gleason, 1995.

Family photo of Mrs. Andrew Gleason (Edith) with her sons in the early 1930s.
Front row (left to right) David James, John.
Back row (left to right) William, Edith, Roger, Lawrence.

Andrew Gleason in smoking jacket.

Portrait of Eleanor Gleason.

19 Kate Gleason to White, June 30, 1923.

20 Ibid.

21 Kate Gleason, from aboard S. S. Pittsburg, July 31, 1923.

22 In his will dated December 23, 1927, Clemenceau, who died on November 24, 1929, bequeathed to Henry Selden Bacon "the sleeping Buddha, in bronze, which is on the table in my bedroom."

23 Kate Gleason to White, from aboard the S. S. President Van Buren, n.d.

24 "Woman, Talented in Many Professions, in City," *San Francisco Chronicle*, March 5, 1924, 10.

CHAPTER 10

The Lucky Rover

WITH HER NEW REAL ESTATE ENTERPRISE IN MOTION, Kate headed back to Europe in June 1924. She sailed aboard the *Scythia* from New York on Thursday, June 19, accompanied by Lillian Gilbreth and other engineering friends. Lillian was heading to a first-of-its-kind International Congress of Management in Prague, Czechoslovakia, scheduled for the end of July. Frank Gilbreth had been instrumental in launching this venture with ASME, the Masaryk Academy in Czechoslovakia, and three other American engineering societies. On the way, he and Lillian planned to attend the World Power Conference in London, where Kate would be ASME's representative.

Five days before they were to sail, Frank Gilbreth suffered a heart attack and died in a Montclair, New Jersey, phone booth while talking to Lillian. Despite the trauma of her husband's death, Lillian realized that she was suddenly the sole support of her eleven living children and felt that she had to continue the work she had done with Frank and establish herself in the international industrial engineering community. Even though several of her children were sick with measles and chickenpox, Lillian believed they were well cared for and decided to carry on in Frank's place as a delegate to the Prague meeting. Elizabeth (Libby) Sanders, a neighbor and Smith College friend of her eldest daughter, accompanied her as a companion and helper, and both women kept diaries of the trip.

Aboard the *Scythia*, Lillian immediately established a daily four o'clock tea in the Garden Lounge for "perhaps a dozen of us—that is the best part of the day!"[1] At the end of the first day, Libby recorded in her diary, "a fascinating person, Miss Gleason, held the stage," charming the crowd. "She is the jolliest and most adorable person," Libby added, "and knows all there is to know about building houses."[2] Although the next day was rough, and many aboard ship were feeling sick, Libby carried on with her new duties on the ship's dance and entertainment committees. The organizer referred to her as "the dignified lady from Smith," which amused her tremendously, as did the whirlwind of shipboard social life. There were quite a few young people on board, to Libby's delight, and she also cherished the company of her older companions. On the fourth day, she recorded, she "went to Mrs. Gilbreth's tea and had lots of fun with Mr. Rushmore and Miss Gleason, both of whom I love."[3] As the deck sports and card tournaments came to a close, Lillian wrote "I have been a peaceful onlooker. All I really do is to get together the little crowd of 'best friends' for tea every afternoon at four."[4]

On June 29, Kate, Lillian, and Libby landed in Liverpool and spent the first hour ashore hunting for lost luggage, then rode to London in a special railway car that had been reserved for their group. The trio took a bus ride to Wembley, where they strolled with other companions, and attended the World Power Conference, where the Prince of Wales delivered the opening address, a thrilling experience. The next day, Kate took Libby to visit some business friends, and one of them loaned them his car and chauffeur to tour the Tower of London, St. Paul's, and Buckingham Palace, followed by lunch at the London Tavern. The next day, Kate and Libby met at the Hotel Cecil, where Kate was staying, and took a Cook's Tour of Windsor Castle, followed by dinner and a sail on the Thames. They were enjoying each other's company, and Kate offered Libby "a marvelous job in Rochester for a month after we get back," Libby reported, an opportunity that delighted her; "I surely hope it works out," she wrote in her diary.[5]

The next day, Kate met Lillian at ten, and together they visited Atalanta, a little "engineer's works" managed entirely by women. They lunched together before purchasing their tickets home on the Holland American Line. On July 4, Kate brought Libby to a luncheon celebration of the American holiday arranged for them at the factory of her business acquaintances, and from July 5 to July 7, she took Libby along on a three-day motor tour through the English countryside. She had invited the Gilbreths to come with her before Frank's death; Libby took Frank's place and joined Lillian, along with ASME member Louis Marburg and his wife, who had also been in London to attend the Power Conference. The group set off in a limousine that Kate had arranged and spent two nights at the Leicester Arms in Penshurst, where they walked to nearby villages, attended Church of England services, rested, and relaxed.

On July 7, they toured Canterbury before dropping Kate off at Dover, where she boarded a boat to cross the channel to France and continue to Paris. The others returned to London, and eventually to Prague for the opening of the Congress on July 20. Returning to Paris after the meeting, they were joined by Kate and boarded the boat train for Boulogne on the sixth of August. Later, steaming home aboard the *S.S. Volendam*, Kate explained to Lillian how she built her houses so economically. "It is so interesting," Lillian noted in her journal; "I want to learn as much of it as I can."[6] Kate loaned Lillian books, since she had run out of reading material, but they were mostly books of adventure, and Lillian didn't find them very interesting.

Kate, apparently, was watching her weight again. On August 9, Lillian wrote, "Libby, Miss Gleason and I are having a great deal of fun over our waiters. They are so disturbed because Miss Gleason is dieting, and takes only tea, orange juice or lemonade for a meal. They just simply can't understand it, and are sure it is because she is dissatisfied either with the food or the service."[7]

The days aboard the *Volendam* were unexciting and restful. Libby and Kate played cards, often Russian Bank and Auction Bridge, while Lillian, who had neither talent nor interest in card games, was always happy to walk the deck and hear more about Kate's building methods. Lillian and Kate at last parted company on August 17 when they reached New York, while Libby joined Kate in East Rochester, where she spent the remainder of her summer vacation working for her new mentor. Libby brought along a friend who was also beginning her junior year that fall, and the two girls spent a month working at Concrest. Kate put them up in furnished apartments at Genundawah Court, next to the building site, and the girls worked from 7:30 a.m. to 7:30 p.m., hours of their own choosing. They quickly realized that overalls were in order for their construction chores, which consisted of installing panes of glass in French windows, building doors for a dozen closets, painting floors, staining woodwork and furniture, decorating fireplaces, and putting finishing touches on wiring and heating systems. Their overalls slowly acquired many layers of paint and stain, attesting to their enthusiasm that summer, and they returned to their studies with some reluctance and considerable knowledge of homebuilding.

When Libby graduated from Smith College, she took a job as Kate's private secretary. She thought Kate was the "Empress of the Universe." Libby had been only seven years old when her mother died, and she had never been very close to her stepmother. Libby's sons grew up with the words, "What would Miss Gleason have thought of that?"[8] ringing in their ears each time they made a misstep. Kate, for her part, was very fond of Libby;

perhaps she saw in the young woman the daughter that she never had. It was a congenial relationship.

In early 1925, Kate was back in Paris with an interest in finding a way to help France recuperate from the war. She was well aware of the many wealthy Americans who were working to assist in that endeavor, including Anne Tracy Morgan, the daughter of J. P. Morgan, and her friend Anne Murray Dike, who established the American Friends for Devastated France, providing relief and reconstruction assistance to bomb-blasted villages in the Aisne region. They had established headquarters in Chateau Bléran-court, near Soissons in the north of France, and Morgan tirelessly mobilized hundreds of American women to raise funds for recovery efforts.

Kate undoubtedly had such a cause in mind when she searched Paris newspaper advertisements for an absorbing new project. She discovered that a twelfth-century tower and battlements were for sale in the village of Septmonts, not far from Blérancourt. The contact listed in the newspaper was a wine merchant acting on behalf of a village notary. He, in turn, represented an elderly couple caring for five grandchildren, orphaned by the war, and they were selling the property to create a fund for the children's education.

During World War I, some of the heaviest fighting was in northern France. Septmonts, nestled in a valley northeast of Paris, ten kilometers from Soissons, is in an area long notorious for bloody battles. Its chateau and fortifications were besieged over the centuries, sacked during the Revolution, and bombarded during the world wars. In July 1918, the First Division of the American Expeditionary Forces had waged a fierce battle to dislodge the Germans from the area. The Americans won the four-day battle but suffered devastating casualties; nearly ten thousand American soldiers were killed or wounded. Seven years later, Septmonts had not recovered from the devastation.

Kate was captivated by the challenge of restoring the tower and helping to rehabilitate the town. Septmont's twelfth-century tower led to battlements protecting the chateau, and its most enchanting structure was a forty-three-meter high castle keep, or *donjon* in French. The fairytale tower has a commanding presence, observable from the outskirts of the village. Before Kate sailed for New York in April, 1925, she was the new owner of parts of the "historic pile," as a Rochester newspaper described her French property.[9]

Although neither the donjon nor the chateau were part of Kate's purchases, her holdings did include a square tower, two stone houses, a bicycle shop, two gardens, a wine shop with a license to dispense spirits, and a recreation hall. Subterranean passages led to various parts of the complex, one surfacing half a mile away. There were channels from the tower's upper balcony, down which defenders once poured hot lead and pitch on attacking armies. The property also came with the title *Chatelaine de la Tour de Septmonts*—all for the sum of approximately three thousand four hundred U.S. dollars. One can imagine with what merriment Kate assumed her new honorific.

It was the stuff of dreams for the new chatelaine; she had always been addicted to stories of bravado and enchanted by towers. One of Kate's stone houses shared a wall with the chateau, fronting on the village square, and this is where she chose to live, after making renovations. Kate had a remarkable eye for design and turned a rather depressing piece of property into a graceful and arrestingly attractive home.

After returning to Rochester for two months, Kate and her sister, Eleanor, spent the summer and winter of 1926 in her new French home, entertaining a stream of visitors. Kate, at sixty years of age, was still thrilled by the thought of living in a tower, and she began planning how to turn her square tower into a dwelling. The older part of the structure was under the protection of the Societé des Beaux Arts, so she confined her remodeling to its newer sections. Her romanticism was only slightly tempered by age; since the tower's stone steps were quite steep, with well-worn dips in the center of each tread, Kate bowed to realities and installed an elevator of her own design.[10]

Before she entered the life of the village, no American dollars had ever found their way to Septmonts. Kate, however, employed the townspeople in restoring her buildings and attempted to introduce some forms of industry. Observing that French turkeys were smaller than those found in the United States, she personally escorted a brood of fowl from America to France with the aim of improving the stock on a neighbor's farm. Kate also encouraged the cultivation of mushrooms in the region by bringing in fertilizer from the nearby Chantilly racetrack; it amused her to promote the results as "mushrooms manured by champions."[11]

In 1926, Kate gave the village some radio apparatus for the town hall, and she had previously given the local priest a two-horsepower Citroen motorcar so that he could more easily visit his parishioners. It was common knowledge, however, that Abbé Boidoux would have preferred receiving two real horses. Since radio equipment was not on villagers' wish lists either, a deputation of townspeople, led by the priest and the mayor, begged to be allowed to trade the radio in for a new bell for the parish church. Music emanating from a radio could never compete with the rich tolls of their beloved bell, which had suffered extreme damage during the war. Kate agreed without hesitating, and the old bell was recast.

On December 19, 1926, parishioners and town officials from all the neighboring villages assembled in the Septmonts church to baptize the restored bell, which was covered in delicate French lace for the occasion. At the end of the formal blessing, Kate and Abbé Boidoux pulled on a white satin ribbon, ringing the bell. As its peals rang out, tears streamed down the cheeks of onlookers who had lived their whole lives within earshot of its mellow tones. A grand celebration followed, complete with a band concert, and many of Kate's friends were there for the festivities, including Mr. Sedley Peck, an officer of the American Legion, and Dr. and Mrs. Henry Durand, old Rochester friends who were living in Paris.

That evening, Kate entertained thirty visiting dignitaries at her home, serving an elaborate meal that began with consommé, followed by langouste, deer filet, roast stuffed turkey, salad, cheese, fruit, meringue, champagne, wine, and liqueur. At dessert, she rose and spoke in French to all of her guests, declaring her deep respect for France, which had suffered so much during the war, and expressing her pleasure in undertaking restoration projects to benefit the citizens of her adopted town. To this day, the front pew in the Septmonts church bears a nameplate with Kate's name on it.

In 1928, Kate transformed the wine shop she had acquired in Septmonts and turned it into a motion picture theater and library, known as "Bibliotheque de Septmonts," which she furnished herself with three thousand books. She also turned part of the building into a children's room, where local youngsters were entertained with plays and games on Thursday afternoons. She dedicated the remodeled building, on Armistice Day that year, as a memorial to the American First Division.

Kate took on a new restoration project in 1929, purchasing Chateau Bucy-le-Long, near Soissons, with its ten-acre park, for two thousand dollars. The chateau, used as a hospital during World War I, had housed many injured Americans, and Kate hoped that one of her friends would take over the property and do for Bucy-le-Long what she was doing in Septmonts.[12] When her old friend and business associate from East Rochester, Harry Eyer, became seriously ill, he asked Kate to dispense a large sum of money on his behalf to aid the people in France, whom he had come to love during the war, and Kate used some of his funds to give gymnasium equipment to a school in the nearby town of Berzy-le-Sec.

Kate had long believed that "a bold front, determination, and the willingness to risk more than the crowd, plus some common sense, and hard work, win out."[13] In Septmonts, that attitude won her the gratitude of the local people. When Mr. Rundschauer of Cornell University asked Kate what the major industry was in the village, she honestly replied, "Me."[14] By 1928, her reputation for generosity was spreading far beyond Septmonts. That year, Kate laughingly told an interviewer that she had been called "the most popular woman in France."[15]

Alongside her scrapbooks, she kept a Croix de Guerre—the symbol of a grateful French government—in its original box. Kate shared this honor with other philanthropic American women, including Anne Tracy Morgan and Edith Wharton. Captain Hardwin Werth, a retired French Army officer, personally witnessed France's outpouring of affection and appreciation for the lady engineer from Rochester at a parade in which Kate was the star attraction. The streets were jammed with people, he recalled, and as Kate rode by, waving in the back of an open morotcar, the crowds cheered, "Vive la Gleason! Vive la Gleason!"[16]

During the latter half of the 1920s Kate was, as she put it, "a lucky rover over the face of the earth,"[17] splitting her time between Rochester, Sausalito, and France. Her attention had wandered far from East Rochester, and she had begun to arrange her affairs so that she would have to spend less time in New York State. She entrusted her East Rochester tracts to her managing director, Marie Weiland, who considered herself fortunate, since the job enabled her to meet Thomas Edison, Clarence Darrow, George Eastman, and even to drive Henry Ford to the railway station in her Hupmobile.[18]

Despite her high hopes for Concrest, Kate had sold just sixteen of her homes by 1926, and she was seeking a buyer for the forty-one that she had rented out. Two young Rochester men agreed to purchase the homes, along with the clubhouse at Genundawah Country Club, which they wanted to turn into an apartment house. They remodeled and painted throughout the summer but for some reason the property reverted to Kate at the end of 1927. Two years later, she donated all of her East Rochester properties, including seventy-two homes and twenty lots, to what is now the Rochester Institute of Technology, where her family had many ties. In his earliest years in Rochester, William Gleason took courses at its predecessor institution, the Mechanics Institute, as did all of his children. Eleanor served as the school's first librarian, Jim was a member of the board of trustees from 1899 to 1964, and, over the years, many Gleason employees came to the company through RIT's apprentice program.[19]

Kate was still a curiosity in engineering circles. More than a decade after she became a member of ASME, she was still the only woman in the organization. This was not an accident, as Lillian Gilbreth learned. ASME was the organization that Frank Gilbreth had prized most highly, and Lillian, determined to carry on the work she had pioneered with her husband, discreetly asked men on the membership committee about submitting an application. "There is an impression on the part of the Membership Committee," they responded, "that the admission of lady members has not been an entire success and a disposition not to encourage it in the future."[20] Lillian persevered, however, and was admitted to the organization in July 1926, soon becoming one of its most valued members.

Back in Rochester at the beginning of 1927, Kate was preparing a speech on her East Rochester projects for an ASME meeting on January 12th. Her focus, however, was Sausalito, and, after the meeting, she returned to the Bay Area to begin her new building enterprises. In early 1925, Kate had arranged to purchase six beachfront lots in Old Sausalito, abutting the Fort Baker Reservation on the south. She intended to build a home for herself on this property, named "Roca Cuadrada," or, "Square Rock." In 1926, she had bought another fifty-four lots in Sausalito from trustees and creditors of the Sausalito Spring Water Company; another nine lots at auction, for one thousand one hundred and fifty dollars, from the trustee of the bankrupt Continental Building and Loan Association; and two more lots at auction from the tax collector of Marin County. Kate was making good use of her experience as receiver-in-bankruptcy. In all, she purchased at least eighty-six verifiable lots in the town, far fewer than the six hundred that she was said to own there.

Kate incorporated much of her property, including seventy-eight lots, under the name of Tierra del Agua Pura, Inc., "Land of Pure Water." Indeed, many of Sausalito's famous springs were located on land now owned by Kate. She excluded from the corporation those lots on which she intended to build her home, on the east side of Alexander Street, next to Fort Baker, as well as the property at Atwood and North streets, where she lived when she was in town. She gave one share of capital stock in the new corporation to each of four Rochester friends, reserving one hundred and ninety-six shares for herself, and later gave five shares to Robert Miller, who was transacting all of her Sausalito real estate business.

To launch her new development project, Kate hosted a beach luncheon on April 2, 1927, at the Alexander Street site, entertaining her friends with a bonfire and chicken a la king. Her guests, in addition to Libby Sanders, included Robert Miller and his wife, her old friend Archbishop Hanna, and James McLaughlin, who had called her to San Francisco three years earlier during Berkeley's rebuilding. Two days after the lunch, Kate was granted construction permits and began her Sausalito homebuilding in earnest, though she did not linger there to personally oversee the work.

The next year, the Bay Area's dream of a bridge connecting Marin County and San Francisco moved closer to reality with the formation of the Golden Gate Bridge and Highway District. Although there was ferry and train service to Marin from San Francisco, they were relatively slow forms of travel and effectively isolated Marin County from the city. Ordinarily, Kate would have considered the planned Golden Gate Bridge a bonanza, since it would make her properties more accessible and thus more valuable. Unfortunately, the planned highway to and from the bridge on the north side went right through many of her properties, which the state acquired by eminent domain. Much of her other real estate, located next to the new highway, would be impacted by the construction process.

As a result, in 1929 Kate made a gift of her shares in Tierra del Agua Pura, Inc., comprising almost all of her Sausalito property, to St. Vincent's School for Boys, an industrial school in San Rafael, California, under the tutelage of Archbishop Hanna. Although Kate had lost some of her property to the "Redwood Highway," the remainder of her land was a magnificent gift whose value, in today's terms, is still incalculable. In appreciation, Archbishop Hanna named St. Vincent's new building "William Gleason Hall." "That would not mean much to Father," Kate remarked in a letter to Jim. "But Dr. Hanna is to deliver the dedication speech and tell the boys what to emulate in Father's life. And once Father told me he would like to have Joseph O'Conner write his obituary, or Dr. Hanna. So he gets his wish."[21]

NOTES

1   Lillian Gilbreth, diary no. 7, 1924, 11. The diaries are in possession of Lawrence Rowland.

2   Elizabeth Sanders, diary no. 1, 1924, 102.

3   Sanders, diary no. 4, 106.

4   Gilbreth, diary no. 8, 12.

5   Sanders diary no. 8, 111.

6   Gilbreth diary no. 49, 86.

7   Ibid., diary no. 50, 88.

8   Lawrence Rowland (son of Elizabeth Sanders), in discussion with the author, March 2003.

9   "Twelfth Century French Tower Is Purchased by Kate Gleason: Historic Pile Will Be Kept as Memorial," *Rochester Times Union*, April 27, 1925, 16.

10   The elevator is long gone, and today's visitor must make the climb in the damp dimness to explore the tower's upper levels. Kate's second stone house was destroyed during World War II.

11   Rundschauer, obituary of Kate Gleason, 178.

12   The chateau was still in Kate's possession at the time of her death.

13   Bennett, "Kate Gleason's Adventures," 168.

14   Rundschauer, obituary, 178.

15   *Rochester Democrat and Chronicle*, July 28, 1928.

16   Captain Hardwin Werth (French teacher, Allendale School, Pittsford, New York), in discussion with James S. Gleason, 1940s.

17   Kate Gleason to Mrs. Andrew White, April 23, 1923.

18   "Marie Weiland: Buildings Manager," *Rochester Times-Union*, August 18, 1983, 5B.

19   In 1939, James E. Gleason succeeded Henry Lomb of the Bausch and Lomb Company as president of the board of trustees of Rochester Institute of Technology. He served in that capacity for the next twenty-five years before retiring from the board in 1964.

20   Graham, *Managing on Her Own*, 101.

21   Kate Gleason to James Gleason, December 21, 1929. Joseph O'Conner was the editor of the *Rochester Post Express* and considered to be the dean of Rochester journalists.

CHAPTER 11

A Woman of Consequence

K ATE'S ENTERPRISES DID NOT JUST REACH from coast to coast and
across the Atlantic. In 1927, she looked south as well, to the seaport
town of Beaufort, South Carolina. Libby, a Beaufort native, had been telling Kate stories about the city for three years. Founded in 1711, Beaufort
had once been a cultural center and commercial hub for indigo, cotton, and
rice. Located on the Beaufort River and cooled by its salty breezes, the city
was known as the "Newport of the South" in the nineteenth century, and
its rich wildlife drew hunters and fishermen from the North. By the 1920s,
however, Beaufort was economically depressed, and Libby hoped Kate
would be a rescuer for her Sea Island paradise.

Kate was intrigued by the possibilities. In June 1927, she and Libby
made a whirlwind trip down to Beaufort, and Kate spent ten hours touring
the area—enough time for her to make up her mind to invest in development projects in the region. She also realized it would be an ideal place for
her widowed stepmother to winter. Kate immediately set up a corporation
named "Sea Island, Inc.," with a capital of a hundred fifty shares and a par
value of one hundred dollars. The next month, she returned to Beaufort,
this time for four days, and brought her stepmother, Margaret Gleason. As
in Sausalito and Septmonts, Kate wanted to secure lodgings for herself before launching a larger project, and she was hatching big plans for Beaufort.

Her company, Sea Island, Inc., purchased riverfront property on New
and Bay Streets for twelve thousand dollars. Located in the heart of town,
with a grand view of Beaufort Bay, the site featured a dwelling that had
been the home of Henry W. de Saussure, director of the United States Mint
under George Washington. The home had been built in 1795, the same
year de Saussure minted the first "gold eagle" coins. Kate discovered that

the frame of the old house was joined together with wooden pegs and that the foundation was cemented with tabby, a substance that was made with oyster shells. She at once began to restore the antebellum home, repairing the façade with yellow stucco and fashioning the attic rooms into an apartment for her own use. On three sides of Kate's cheery sitting room—full of glazed chintz upholstery and bright curtains—large windows looked out on wide vistas of islands and marshes along the Beaufort River.

Kate saw Beaufort as an ideal location for vacationers and retirees, so she designed and built a thirty-one room hotel and tavern that was connected to her own house by a pergola. All of the hotel's bedrooms had "en suite" baths, a rare amenity, and the ceilings and paneling were made of polished native pecky-cypress, half as costly as plaster. Beaufortonians could be excused for referring to her as "Concrete Kate," since she constructed the new building of poured concrete, and they may have wondered about the unusual aesthetics of her design. The hotel's stairway was enclosed in a round tower with a conical roof, like one she had built in Marigold Gardens in East Rochester. Although the tower might have had practical value and was no doubt sentimental for Kate, it was, aesthetically, somewhat jarring.

By August 1927, Sea Islands, Inc. had also acquired several parcels on Lady's Island across the river from Beaufort, and in October, the *Beaufort Gazette* announced that Kate had purchased Dataw Island, a six-mile journey from Beaufort by flat-bottomed boat.[1] The Sams family, who had lived on Dataw from 1690 until the Civil War, had been successful in growing indigo and long cotton on the island and had planted extensive orchards of plums, figs, apples and pears. They had also brought young orange trees from France and were believed to have been the originators of America's orange industry. When Kate acquired the island for fifteen thousand dollars, however, there was very little left of their island kingdom—only the ruins of the Sams's mansion, a church and graveyard, and a handful of black families who still eked out a living on Dataw. The island was another casualty awaiting rescue.

Kate's first thought was to turn it into a huge turkey farm that could employ the local population. To set the project in motion, she planned to bring out a friend, Madame Claire Laur—who was successfully promoting the turkey industry on her farm near Septmonts—and two French families to teach South Carolinians how to raise commercial flocks. It is unclear whether the French turkey experts ever made it to Dataw, but there is a Gleason family story about a carload of turkeys, purchased by Kate, that went awry. The train car was mistakenly routed to Beaufort, North Carolina, instead of Beaufort, South Carolina, and the turkeys languished on a railroad siding over the Fourth of July weekend, perishing from the heat. Kate's Dataw turkey initiative never got off the ground, and the turkeys that were there were said to have ended up in stew-pots all over the island.

As one reporter noted, however, Kate was never without several strings to her bow.[2] As part of her "Southern Reclamation Project," she wanted to commercialize the rich natural assets of the coastal areas. Kate developed plans to rehabilitate the entire Beaufort region by urging industries and sportsmen to consider its possibilities. Land in the vicinity of Beaufort, she argued, was particularly suited to the production of vegetables and fruit and could be bought for thirty dollars an acre, one-tenth its cost before the Civil War. "I am anxious," she explained, "to induce some cannery company to establish a branch factory in the district so that the present growers, and others who will be induced to come, may have a market for their products. There is an abundance of labor available and electric power is cheap so that conditions for manufacturing are ideal."[3]

At the same time Kate was developing Septmonts and Sausalito, she had found, in Beaufort, another grand civic project and opportunity. She was, she said, "what the world calls a success. I have done what I set out to do, and much more." Now, she added, "I want to go where people want me, and work for them. That is why I went to France," she explained, and "why I am working in Beaufort."[4]

In April 1928, Helen Christine Bennett, a well-known magazine writer, visited Kate at Clones in Rochester to interview her for the *American Magazine*. The article, titled "Kate Gleason's Adventures in a Man's Job," appeared the following October and generated considerable interest in Kate, as well as several proposals of marriage. Never shy about publicity, Kate must have delighted in all the attention. Bennett and her husband became good friends of Kate's, visiting her in Beaufort and ultimately settling there.

That same year, Herbert Hoover, Secretary of Commerce, ran for president on the Republican ticket. He and his wife, Lou Henry Hoover, were both engineers, and Kate and Lillian Gilbreth—fellow engineers and fellow Republicans—contributed their names, energy, and money to his successful campaign. Lillian even served as president of the Woman's Branch of the Engineers' National Hoover Committee; Kate was one of the group's seventeen honorary vice-presidents, and its officers included Mrs. Henry Ford and Mrs. Thomas Edison. Many of the members—including Kate, Lillian, and Lou Henry Hoover—also founded the Engineering Women's Club in

1928; its first headquarters in New York City became an international center for women engineers and the wives of engineering professionals.

Kate was still actively involved in ASME, and on May 13, 1929, when its members convened a four-day spring meeting in Rochester, Kate lavishly entertained five hundred participants at Clones. She had transformed her garage into a motion-picture theater and card room for the evening, constructed a separate building for a men's lounge, and organized sporting activities on her spacious lawns. Dinner was prepared and served by the Food Administration Department of the Rochester Mechanics Institute. After dinner, guests sang songs with piano accompaniment, listened to a male quartet, watched movies, and played cards, followed by dancing. Kate always knew how to throw a party.

She was, according to Bennett, "a handsome, stalwart, robust woman with magnificent white hair, a woman of great gifts and great achievements who lives heartily and candidly; a woman whose abundant vitality and rollicking humor is finely matched by her great mental ability."[5] Kate's rollicking humor, however, was not always well-received. On the first evening of the ASME conference, she was called upon to say a few words and, to the shock of many in the audience, made a joke that would have been more appropriate on the factory floor. "I will make this brief," Kate began, "like the man who wrote 'I have your letter of the eighth before me. It will soon be behind me.'"[6] Some thought Kate did not know the meaning of her words, but she understood the joke perfectly well and thought it funny. Kate's morals may have been prim, but she was free with her tongue.

She was still rocketing nonstop between Rochester, France, and South Carolina. That summer in Septmonts, she entertained her sisters-in-law, Edith and Miriam Gleason, as well as Edith's eighteen-year old daughter, Ellen. Jim had begged Edith to take his wife with her to Europe, since Miriam had few friends and Jim certainly didn't want to travel with her. Edith reluctantly agreed to include her sister-in-law out of pity for Jim. By this time, he and Kate were on a friendlier footing. His wife and son spent time with Kate in South Carolina, and Kate hosted Jim and some of his business friends on a vacation in Beaufort. At this point, Andrew was the only member of the Gleason family who was still hostile to Kate. Her estrangement from Andrew, however, did not extend to her relationships at the Gleason Works; she invited employees to picnics at Clones and always held their interests close to her heart.

In 1929, Kate, an aficionado of pre-Revolutionary art, was the guest of honor at the Art Club of Washington, D.C., where her old friend Emma

Michel Schlick and her husband, who were both artists, had resided for years. That fall, Kate also returned to the Far East, traveling with Lillian Gilbreth as delegates to the World Engineering Congress in Tokyo. They first headed west with friends and colleagues across the United States, boarding a special train from New York City to San Francisco and making stops at the Grand Canyon and Los Angeles. So many participants were planning to attend the Tokyo conference that they filled two steamers, which transported them across the Pacific to Yokohama by way of Honolulu.

Kate finally arrived in Japan on October 27, 1929, a day after a violent typhoon and three days after the New York stock market crash that triggered the Great Depression. Nearly seven thousand miles across the globe, Kate and the other delegates were a world away from the disastrous event, enjoying exquisite entertainments including an Imperial garden party, visits to the Imperial and Kabuki theaters, and demonstrations of Japanese flower arranging and bonsai gardening, as well as trips to Nikko and Hakone to view ancient temples and mountain scenery. After the conference, many attendees visited factories, schools, and engineering works in other parts of Japan, as well as Korea, Formosa, and China, with free rail passes distributed by the congress.

Back in Beaufort at the end of the year after her exhilarating and exhausting trip, Kate had received welcome financial news. The Gleason Works was in high gear under Jim's able management, and Kate gratefully wrote to him in Rochester, admitting that she was now more dependent on her income from the family business. "Anyone would enjoy such an immense dividend—I certainly do," she wrote. "A few years ago my income outside the Gleason Works was bigger than the Gleason Wks dividends but it is different now."[7]

Still, Kate's feverish buying and building in South Carolina was undiminished, and she extended her generosity to the Beaufort community. When the local People's State Bank was facing bankruptcy, she deposited a check for twenty-five thousand dollars in the institution, the largest in memory up to that point, enabling it to avoid insolvency.[8] Kate also extended her help to a Beaufort family. In a postcard from South Carolina, she told the Walters in East Rochester that she had "a little girl here. She was #9," Kate wrote, "and her father did not want her. I am financing the mother so she does not have to give the baby away."[9] The Walters, especially, could appreciate that generous gesture.

She was plunging ahead with her ambitious plans for rescuing Beaufort. In February 1930, she completed the hotel attached to her residence, named it the "Gold Eagle Tavern" in honor of de Saussure, and leased it to seasoned hoteliers, Mr. and Mrs. L. E. Wilder. The couple managed it brilliantly, earning the tavern a national reputation for culinary excellence. Two hundred people, including many visitors from the North, celebrated its opening in March 1930. Local residents often dined there on Sundays along with newlyweds, actors, writers, and artists—including Clark Gable, Somerset Maugham, and Adlai Stevenson—who frequented the Gold Eagle Tavern as hotel guests. Kate maintained an apartment there for the rest of her life.

The first of her plans for promoting Beaufort was now fully realized, but Kate had many more. She purchased land and planned a resort across the river on Lady's Island, where artists and writers could vacation at a modest cost, although the property eventually became a broader development called Colony Gardens. In 1930, Kate also financed construction of a causeway between St. Helena and Polowana Islands and another from Polowana to Dataw. The project employed dozens of island residents and was a lifeline that enabled many of them to save their homes during the Depression. To oversee construction, she hired H. Reeves Sams, who had been born on Dataw and was a descendant of the island's owners. "Since he has taken charge," she noted, "the old people who were born on Dataw are flocking back so as to be near him."[10]

With all her social activities in the Beaufort area, Kate thought it would be nice to have a yacht for entertaining and motoring around the Sea Islands off the Atlantic coast. She bought a one-month subscription to a Jacksonville, Florida, newspaper and began scanning advertisements. Kate found a yacht that was used as a rum-runner between Cuba and Jacksonville, before it was captured by the Coast Guard and sold at auction. She bought it for a thousand dollars, named it "Ellen of the Isles" after her mother, and later discovered, to her immense delight, that the boat had originally cost twenty-eight thousand dollars. Kate always loved a good bargain.

In her sixties, she often invited friends aboard for a day's sail and usually invited a young couple, Tony Agostinelli and his wife, whenever she entertained visitors from Rochester. Tony's uncle, Alphonso Agostinelli, was Kate's trusted foreman; he had worked for her in East Rochester and was now overseeing her building projects in Beaufort. Since Kate was creating more work than Alphonso could handle alone, she hired his nephew, Tony, to help out. He and his wife drove Kate's new station wagon down from Rochester, and she gave them one of her new houses to live in on Lady's Island. Since Tony's new bride was often lonely and homesick, Kate let them take her car to go to the movies and dances, and did whatever she could to make life more enjoyable for her.

Kate also sailed with family members, especially her sister, Eleanor. Eleanor had made a life for herself outside the family business—after the war, she had set up the Dewey Decimal System in the libraries of the U.S. Virgin Islands—but she was, more and more, a helper and co-host for her aging sister. It was inevitable that after a luncheon aboard the yacht, Kate would retire to her cabin for a nap, leaving Eleanor to entertain all of her guests. Eleanor also remembered, with some annoyance, the time Kate invited eight people for dinner, and then forgot all about it and went out for the evening. Eleanor was home alone when she answered the doorbell and found eight hungry people waiting outside. Somehow, she managed to put food on the table, but it was a shock, and no doubt insulting to her guests that Kate had forgotten about them.

Eleanor was also her sister's traveling companion. In April 1931, the two of them sailed for Europe, bringing the station wagon since they intended to take an extended driving trip through Normandy, England, Scotland, and Ireland.[11] Years later, Eleanor rolled her eyes and held her head in her hands when she recalled the journey. Eleanor liked to sleep in, but Kate was an early riser who greeted the break of day with abundant energy. By the time Eleanor forced herself out of bed, Kate had already returned from a brisk walk around whatever village they were in and knew all about the history of the place. The most unforgettable part of the trip for Eleanor, however, was the portable potty that Kate insisted on carrying in the back of the car.

Kate loved the trip, especially their visit to Cleary Castle, an ancient, romantic fortress in her ancestral homeland. "We ought to see some Fairies in the castle to-night," she wrote her nephew Emmet on Midsummer Day. The castle, she told him, was "Norman about 13th or 14th century and it had a moat and portcullis and drawbridge and everything. You ought to buy it," she suggested, "and fix it up."[12] Emmet did not buy it, and, more surprisingly, neither did Kate. Perhaps she realized that she already had enough castles and construction projects to deal with.

Kate once told an interviewer that, "I safeguard my health jealously, for I want to live to be a hundred years old and to work until the last minute."[13] At sixty-six, however, she wrote Lillian Gilbreth, "I have more than enough work now to take the rest of my life."[14] She had no intention of slowing down. After returning to Beaufort in the fall of 1931, Kate sailed again for France in early May and remained in Septmonts until the end of July. The next year, she continued her itinerant lifestyle, spending the summer in Septmonts and the fall overseeing her Colony Gardens project in South Carolina.

At the end of 1932, she traveled to Rochester for the Christmas holidays. Although the city had experienced one of the warmest Decembers on record, Kate came down with pneumonia and was admitted to Rochester's Genesee Hospital on January 3. At 11:06 in the morning of Monday, January 9, 1933, she died at age sixty-seven, thirty-three years shy of the century mark she had hoped to attain.

Kate's funeral was held two days later at Clones.[15] One end of her spacious living room was banked with flowers, a string trio accompanied the service, and twenty-four employees of the Gleason Works served as honorary pallbearers. Condolences arrived from friends and associates around the world. Beaufort's mayor, W. R. Bristol, wired that it was "a great shock to this community to learn the sad news of the passing of Miss Kate Gleason. Beaufort has lost one of its warmest friends and most valuable citizens."[16] The *Charleston News and Courier* noted in Kate's obituary that she was "known to have done a great deal of relief work here, most of it secretly to all save a few friends. She did not contribute heavily to organized charity in this section, but found many ways to help those in need."[17]

Announcements of Kate's death appeared in professional journals, the *New York Times*,[18] *Time* magazine,[19] and many other periodicals across the country. The *Rochester Times-Union* declared that "Miss Gleason had a full, varied and most useful life ... as a pioneer in new fields of women's work."[20] The *Cornell Alumni News* reflected that "Kate Gleason has been called eccentric. That must be because it is considered eccentric to know exactly what you want to do, and to do it without regard for the taboos of the weak, the apathetic, and the timid."[21]

The *American Machinist* magazine predicted that

> News of the death of Kate Gleason—will cause many an old timer in the industry to pause and recall her visits to his shop as a sales representative of her father's company in Rochester. Some of them will recollect her first visit which usually resulted in the discomfiture of some smart engineer who undertook to entangle her on gear problems. There is no record of anyone making much progress in that direction.[22]

And, according to the *Cleveland Plain Dealer*,

> American feminists should erect a monument to Kate Gleason, whose recent death in Rochester, N.Y., closed a career which bristled with "firsts" in

fields where women are still rare pioneers. This remarkable woman did not spend much time discussing what women might do in competition with men. She did them.[23]

NOTES

1   N. L. Willet, "Dataw Island, Oranges and Two Brothers of the Sams Family," *Beaufort Gazette*, October 27, 1927, 2. The name of the island was sometimes spelled as Dahtaw, Datah, or Dawtaw, and is currently spelled Dataw.
2   Amy H. Croughton, "Noted Woman Writer Pays Kate Gleason High Tribute," *Rochester Times-Union*, April 19, 1928, 30.
3   Ibid.
4   Bennett, "Kate Gleason's Adventures," 168.
5   Ibid.
6   Ellen Gleason Boone, in discussion with the author, April 1986.
7   Kate Gleason to Jim Gleason, December 21, 1929.
8   Mrs. F. W. Sheper, Jr. (widow of the former cashier of the People's State Bank in Beaufort), in discussion with the author, April 1986.
9   Kate Gleason to Mr. and Mrs. Walter, n.d.
10 Kate Gleason, letter to the editor, *Journal of the American Concrete Institute*, December 1930, 6.
11 In June 1930, Kate had attended the Second World Power Conference in Berlin as ASME's special representative. By fall, she was back in Beaufort, and, after a brief Christmas visit to Rochester, she returned south at the beginning of 1931.
12 Kate Gleason to Emmet Gleason, June 21, 1931.
13 Ross, "Kate Gleason of Rochester," 29.
14 Kate Gleason to Lillian Gilbreth, from Gold Eagle Tavern, December 9, 1931.
15 Kate's funeral was conducted by the Rev. Justin Nixon, D. D., minister of Brick Presbyterian Church. Jim Gleaon's wife and son belonged to Brick Church.
16 *Beaufort Gazette*, January 12, 1933, 1.
17 "Miss Kate Gleason, Beaufort Developer, Dies, Pneumonia," *Charleston News and Courier*, January 10, 1933, 2.
18 "Kate Gleason, Engineer, Dead," *New York Times*, January 10, 1933, 21.1.
19 "Milestones" (obituary of Kate Gleason), *Time* 21, no. 4 (January 23, 1933).
20 "Kate Gleason Dies; Made Name for Self in Business World," *Rochester Times-Union*, January 9, 1933, 1.

21   Rundschauer, "Just Looking Around," obituary of Kate Gleason, 178.
22   *American Machinist*, January 18, 1933, 33d.
23   "A Woman Who Was First," *Cleveland Plain Dealer*, January 19, 1933; repr.,
     Rochester Engineer 11, no. 8, February 1933, 99.

CHAPTER 12

Legacy

KATE DIED A WEALTHY WOMAN; she had earned every cent of her fortune and left an estate appraised at $1,413,881.55. The value of what she had already given away during her lifetime is largely unknowable. Her will was made public on January 14, 1933. She had signed it the previous spring on a hurried trip to Rochester, misspelling names and omitting addresses in her haste, causing considerable confusion for her executors. She left bequests to more than seventy individuals and organizations. Many of those to whom she left money had no need of it, but it was Kate's way of honoring their friendship.

To her sister, Eleanor, she left all of her personal effects, including jewelry, books, automobiles, and furniture, as well as her preferred stock in the Gleason Works, valued at a hundred dollars per share. She also left her a number of mysterious "manuscripts." In a letter to Jim, Kate once referred to a history of the "McDermots and Glesons" that she and Eleanor were writing, and that document may have been among the personal papers that Eleanor elected to burn.

Kate had always warmly remembered her high school history teacher Amelia Brettels, preceptress of the Rochester Free Academy, who had died in 1890. In her will, Kate left one hundred thousand dollars to the City of Rochester to establish a history alcove in its public library as a memorial to Brettels, "to whom," Kate wrote in her will, "I am grateful for the inspiration given by her to me in the study of history."

Kate also left a hundred thousand dollars to "Dr. Lorenzo Kelley—to whom I am grateful for his expert knowledge in the use of radium." In her hurry, Kate wrote down the wrong name; she intended the bequest for Dr. Howard A. Kelley, a surgeon who operated a private hospital in Baltimore,

Maryland. When Kate was ill in 1917, she had been treated by Dr. Kelley on the advice of her sister-in-law, Edith. A cancer specialist with ample private resources, he had no need for the money and considered using it to deliver hospital services to the poor. Kate's mention of radium is a little puzzling. It may have been prompted by a successful fundraising campaign, launched by the *Delineator* magazine, to purchase radium for the Nobel Prize-winning French physicist Marie Curie. Madame Curie had visited the United States in 1921 and 1929, each time receiving a gram of radium, and Kate, who was once called the "Marie Curie of machine tools,"[1] could not have missed the frenzy surrounding Curie's visits. She was certainly a fan: an 8 1/2 x 11-inch photograph of the celebrated scientist is among Kate's effects, inscribed with the words, *"A Mademoiselle Kate Gleason avec le meilleur souvenir de Marie Curie"* ("To Miss Kate Gleason with best regards from Marie Curie").

To her old friend Henry T. Noyes, Kate bequeathed twenty thousand dollars and a forty-nine-acre parcel on the Genesee River, known as River Farm. Noyes, founder of a company called Art in Buttons, was an enlightened employer and civic leader who helped establish the United Charities of Rochester, a forerunner of the United Way. He often visited Kate at her various homes and was one of many individuals Andrew detested. To her friend Archbishop Hanna, Kate left her remaining Sausalito property, Roca Cuadrada. Libby Sanders received Dataw Island, comprising 1,051 acres, together with all of its approaches and adjoining property.[2] In addition, Kate left a thousand dollars to each of five long-time Gleason Works employees and five thousand dollars to Schuyler Earl, the manager and a director of the company. She also bequeathed a thousand dollars to the widow of Stanley Fox, a Gleason employee who had died aboard the *Titanic* on his way home from a sales trip in Europe. She left the interest of a ten-thousand-dollar trust to a former housekeeper, five thousand dollars to each of six friends and the infant son of another, and a thousand dollars each to forty-seven other friends, including Lillian Gilbreth, Helen Christine Bennett, Helen White, and Henry Sharpe of the Brown and Sharpe Company, who used the gift to establish the Kate Gleason Fund at the Providence Engineering Society in Rhode Island.

Kate made an additional bequest of twenty-five thousand dollars to the Rochester Engineering Society, which used the legacy for its endowment fund.[3] She also bequeathed all of her French property to the Paris Post #1 of the American Legion. Because the group was not qualified to accept the gift, Kate's French holdings passed into the remainder of her estate, which was placed in a charitable trust, the Kate Gleason Fund. Her sister, Eleanor, and friends Henry T. Noyes and Carlton T. Bown—whose father succeeded Kate as president of the First National Bank of East Rochester— were its trustees.

So much of Kate's estate consisted of non-liquid assets, including her development projects in Beaufort, that the executor was unable to pay legacies without selling a large block of Gleason stock. Since there was no market for the closely-held corporation, the trustees arranged with the company's directors to transfer two-thirds of Gleason common stock to the firm in return for cash with which to pay bequests. The other third of the common shares remained in the Kate Gleason Fund, and their dividends were paid into the Gleason Welfare Fund to benefit employees, as Kate directed. For the next few years, other members of the Gleason family paid into this fund from their own dividends to provide retirement benefit to employees and aid them with emergency expenses; it was a paternalistic approach taken by a number of companies before President Roosevelt signed the Social Security Act in 1935.

When Kate died, her Colony Gardens project was only partly finished. Eleanor agreed to complete the project, with some reluctance. Kate had envisioned a resort development comprised of apartments, individual homes, and central amenities, including a salt-water pool, badminton and croquet courts, horseback riding, and hiking trails through its extensive pinewoods. Eleanor carried out most of her sister's plans, and the complex opened on July 9, 1933, to wide acclaim. As the years passed, Eleanor warmed to the area and bought a small home of her own in the riverside complex. To run the property, she hired Andrew Gleason's daughter Ellen, who had just graduated Phi Beta Kappa from Smith College. Ellen earned her fifty-dollar-a-month salary with energy and intelligence, and Colony Gardens thrived under her apt management.

During her lifetime, Kate had made many gifts in honor or memory of others, but never for the purpose of glorifying her own name. Her many achievements, however, were an inspiration to women entering the almost exclusively male field of engineering. In recognition of her contributions, Rochester Institute of Technology decided, in 1998, to name its engineering school the Kate Gleason College of Engineering, the first ever named for a woman engineer. Kate, who liked to be "first in any field," would have enjoyed the honor; in this legacy, too, she was a pioneer.

Kate also knew that people who do things cannot escape a certain amount of publicity, and her fame, or notoriety, nearly led to a less flattering memorial. On January 15, 1933—six days after Kate's death, which was reported in newspapers across the country—the distinguished playwright Eugene O'Neill was hard at work on a serious new play titled *The Life of Bessie Bowen*. The play, according to O'Neill's biographer Travis Bogard, "appears to have been based on the career of a woman industrialist from Rochester, New York, named Kate Gleason. Whether she was the source figure of O'Neill's heroine or whether her life provided a coincidental parallelism which O'Neill came to know at the time of her death in 1933 is uncertain. O'Neill's notes, however, contain an account of her remarkable career provided by Saxe Commins, a one-time resident of Rochester."[4]

O'Neill had known Commins since 1915. A Rochester dentist, Commins had often visited his talented cousin David Hochstein, a renowned violinist, in New York City, where he enjoyed spending time with Hochstein's friends, especially Eugene O'Neill.[5] In the summer of 1916, the two cousins—whose aunt was the famous anarchist Emma Goldman—spent time together in Provincetown, Massachusetts. A group of young intellectuals had formed a theater group there called the Provincetown Players, and its first production was O'Neill's play *Bound East for Cardiff*. Commins met another of Hochstein's friends on the Cape that summer—radical journalist John Reed, who later wrote the book *Ten Days That Shook the World,* still considered the best eyewitness account of the Russian Revolution."[6] Two years later, in 1918, Hochstein joined the army and died in France at age twenty-five; he was blown to pieces in the Argonne Forest, not far from Septmonts.

O'Neill and Commins remained friends, and in April 1921, the playwright, who had a mouthful of trouble, came to Commins in Rochester for major dental work. While O'Neill was staying with Commins, he may well have heard stories about Kate Gleason. At the time, she was feverishly building her concrete houses in East Rochester; her careers as an industrialist, receiver-in-bankruptcy, and bank president were legends in Rochester, and her Moorish fantasy of a home was a local marvel.

O'Neill's path dovetailed with Kate's, later, at other points. In 1928, when Kate was restoring Septmonts, O'Neill and the woman he would take as his third wife, Carlotta Monterey, moved to France, where they were soon joined by Saxe Commins and his wife, Dorothy Berliner. Bored with dentistry, newly married, and tired of Rochester, Commins longed to become a writer, and France was a magnet for creative American expats. It is certainly possible that O'Neill and Commins caught wind of Kate's Septmont adventures during their years in France, since she was at the time, she reported, "the most popular woman" in the country.[7] Commins subsequently became O'Neill's editor at Liveright Publishing and later Random House, and their conversations over the years may well have touched on Kate.

There was another connection. In 1931, the O'Neills built their home, Casa Genotta, in Sea Island, Georgia, a hundred miles south of Beaufort, where it is possible they also heard about Kate's extensive building activities in that region. It was from Casa Genotta that Carlotta wrote Saxe Commins on January 15, 1933, just days after Kate's death: "Don't hurry with typing *Ah, Wilderness*, as Gene is hard at work on another serious play."[8] It was *The Life of Bessie Bowen*, about a squat, unattractive woman who enslaves men around her in her pursuit of profits. It would be the linchpin of a cycle of plays about American history, using several generations of one family to tell the story. In *The Life of Bessie Bowen*, which began in the last decade of the nineteenth century, Bessie leaves her father's bicycle shop to promote her husband's newly invented cooling system for automobiles. She becomes a dominant force in the automotive industry, extending her reach and power across the country. "At some point in his imaginative transmutation of the life of Kate Gleason," Bogard explains, O'Neill saw her "as an emblem of all Americans who had put aside spiritual values in a greedy drive to possess material wealth."[9] The historical cycle, he adds, "was a work of astonishing scope and scale," but O'Neill was never able to complete it; he and Carlotta burned the manuscript in 1951. Its destruction, Bogard states, was "one of the greatest losses the drama in any time has sustained."[10]

That Kate inspired such a tour de force attests to the extraordinary power of her own achievements. Bessie Bowen, however, was a dark distortion and a grim caricature of the real Kate. Love for family, especially her father, was always Kate's motivation. When her older brother died young, she knew that her father needed her help, and she accepted that responsibility without hesitation. Throughout her life, Kate eagerly added to her responsibilities—growing her father's company, proving that women were capable of contributing in fields typically considered outside their reach, or using her ingenuity and wealth to rescue failing communities. Her life was as productive and positive as Bessie's was destructive and dark.

Of course, Kate had her faults, as her sister and brothers always attested. Andrew, for the rest of his life, never had a kind word to say about Kate, though his attitude was shaped by his inner torment. She and Jim had been very close in their youth and were friendly again in later years; but there is no doubt that Jim sealed Kate's fate at the Gleason Works. Although Andrew's outburst precipitated Kate's departure, Jim was in charge of the business, and she never would have been ousted without his agreement; when she left the business, Jim was solidly at the helm, unchallenged and unhindered by a bossy and very knowledgeable older sister.

Eleanor, too, had mixed feelings about Kate. After she lost her mother at age fifteen, Kate took responsibility for raising her, but she was an especially strict and overbearing substitute mother. Although Eleanor was plucky and capable, she lived most of her life in the shadow of her older sister, and Kate cast a massive shadow, dramatically sweeping through scenes while Eleanor followed, attending to details and patching up problems in her sister's wake. After Kate died, Eleanor lived the last thirty years of her life in houses that her sister had owned or built. When she burned Kate's papers, it may have been less a housekeeping chore than a final declaration of her independence.

In the end, Kate's family problems can be traced to her outsized talents and appetite for work and challenge. Her strenuous life was as full as her boundless energy could make it. It was shorter than she had hoped, but Kate Gleason—the daughter of hard-working immigrant parents, whose own grandmother held her in low regard and whose country did not allow her to vote until she was fifty-five—had an irrepressible "can do" state of mind. *Possum volo*, the words she lived by, expressed her unshakable self-confidence and drive to prove that a woman could work, achieve, and succeed as well as a man: "I can, if I will."[11]

NOTES

1   Colvin, *60 Years with Men and Machines*, 73.

2   This bequest was appraised at $10,810. When Libby married Richard Rowland in 1936, they tried farming on the island but soon abandoned the idea. Libby bequeathed Dataw to her sons when she died in 1965; many years later, they sold the island to Alcoa Properties who developed it in a way that would likely have met with Kate's approval.

3   It is nearly one hundred years since Kate's contentious admittance to the Rochester Engineering Society in 1916 as the group's first woman member.

4   Bogard, *Contour in Time,* 372.

5   Kraut, *Unfinished Symphony*, 30. On October 20, 1982, at the dedication of six portraits of notable Rochesterians, including Kate Gleason, held at Monroe Community College in Rochester, N.Y., Kraut recalled that Kate Gleason taught her how to play poker.

6   Kraut, *Unfinished Symphony*, 31.

7   *Rochester Democrat & Chronicle,* July 28, 1928.

8   Commins, *Love and Admiration and Respect,* 139

9   Bogard, *Contour in Time,* 374.

10   Ibid.

11   Bennett, "Kate Gleason's Adventures," 175.

Bibliography

Altschuler, Glenn C. *Andrew D. White: Educator, Historian, Diplomat.* Ithaca: Cornell University Press, 1979.

———. *Better than Second Best: Love and Work in the Life of Helen Magill.* Urbana: University of Illinois Press, 1990.

American Concrete Institute. *Proceedings of the Eighteenth Annual Convention.* Vol. 18, 1922.

Barry, Kathleen. *Susan B. Anthony: A Biography.* New York: New York University Press, 1988.

Bartels, Nancy. "The First Lady of Gearing." *Gear Technology,* September/October 1997, 11–17.

Bennett, Helen Christine. "Kate Gleason's Adventures in a Man's Job." *American Magazine,* October 1928, 42–43, 168–75.

Bogard, Travis. *Contour in Time: The Plays of Eugene O'Neill.* Rev. ed. New York: Oxford University Press, 1988.

Canel, Annie, Ruth Oldenziel, and Karin Zachmann. *Crossing Boundaries, Building Bridges: Comparing the History of Women Engineers 1870s–1990s.* London: Routledge, 2003.

Chappell, Eve. "Kate Gleason's Careers." *Woman Citizen,* January 1926, 19–20, 37–38.

Colvin, Fred H. *60 Years with Men and Machines.* New York: McGraw-Hill, 1947.

Commins, Dorothy, ed. *Love and Admiration and Respect: The O'Neill-Commins Correspondence.* Durham: Duke University Press, 1986.

Conable, Charlotte Williams. *Women at Cornell: The Myth of Equal Education.* Ithaca: Cornell University Press, 1977.

Conners, Mary, and Jim Burlingame. *East Rochester, New York: 1897–1997.* Dallas: Taylor Publishing Co., 1997.

Coolidge, Mary Roberts. *Why Women Are So*. New York: Henry Holt, 1912.

Cooper, Polly Wylly, and Betty Wylly Collins. *Images of America: Beaufort*. Charleston: Arcadia Publishing, 2003.

Curie, Eve. *Madame Curie: A Biography*. Translated by Vincent Sheehan. Garden City: Doubleday, Doran, 1937.

Dabbs, Edith M. *Sea Island Diary: A History of St. Helena Island*. Spartanburg: Reprint Co., 1983.

Durand, William F. *Robert Henry Thurston: The Record of a Life of Achievement as Engineer, Educator, and Author*. New York: A.S.M.E., 1929.

Ford, Henry. *My Life and Work*. Garden City: Doubleday, Page & Co., 1923.

Gelderman, Carol. *Henry Ford: The Wayward Capitalist*. New York: St. Martin's Press, 1981.

Gilbreth, Frank B., Jr., and Ernestine Gilbreth Carey. *Belles on Their Toes*. New York: Bantam Books, 1984.

———. *Cheaper by the Dozen*. New York: Harper & Row, 1948

Gilbreth, Frank B., Jr. *Time Out for Happiness*. New York: Thomas Y. Crowell Co., 1970.

Gilbreth, Lillian M. *As I Remember*. Norcross: Engineering & Management Press, 1998.

Gleason, Eleanor. "Untitled" (speech at unveiling of bronze tablet bearing the portrait of her father, William Gleason). Gleason Works, January 29, 1931.

Gleason Works. *Fourscore Years of Bevel Gearing*. Rochester: Gleason Works, 1945.

———. *The Gleason Works: 1865–1950*. Rochester: Gleason Works, 1950.

Gleason, Kate. "How a Woman Builds Houses to Sell at a Profit for $4,000: The Story of Miss Kate Gleason's Housing Enterprise at East Rochester, N.Y., Where a Hundred Fireproof Dwellings are Being Manufactured." *Concrete*, January 1921, 9–14.

Graham, Laurel. *Managing on Her Own: Dr. Lillian Gilbreth and Women's Work in the Interwar Era*. Norcross: Engineering & Management Press, 1998.

Graydon, Nell S. *Tales of Beaufort*. Beaufort: Beaufort Book Shop, Inc, 1963.

Gribble, Richard. *An Archbishop for the People: The Life of Edward J. Hanna*. New York: Paulist Press, 2006.

Harper, Ida Husted. *Life and Work of Susan B. Anthony*, 3 vols. Indianapolis: Hollenbeck Press, 1898–1908.

Harrison, Shirley. *Sylvia Pankhurst: A Crusading Life—1882–1960*. London: Aurum Press, 2003.

Kabelac, Karl S. "Kate Gleason, National Bank President." *Paper Money* 38, no. 3; whole no. 201 (May/June 1999): 67–70.

Kotel, Janet. "The Ms. Factor in ASME." *Mechanical Engineering* 95, no. 7 (July 1973), 9–11.

Kraut, Grace N. *An Unfinished Symphony: The Story of David Hochstein*. Rochester: Lawyers Co-operative Pub. Co., 1980.

Lacey, Robert. *Ford: The Men and the Machine*. Boston: Little, Brown, 1986.

Lancaster, Jane. *Making Time: Lillian Moller Gilbreth—A Life Beyond "Cheaper by the Dozen."* Boston: Northeastern University Press, 2004.

Laxton, Edward. *The Famine Ships: The Irish Exodus to America*. New York: Henry Holt, 1996.

Layne, Margaret E., ed. *Women in Engineering: Pioneers and Trailblazers*. Reston: A.S.C.E., 2009.

Leavitt, Judith A. *American Women Managers and Administrators: A Selective Biographical Dictionary of Twentieth-Century Leaders in Business, Education, and Government*. Westport: Greenwood Press, 1985.

Leland, Mrs. Wilfred C. *Master of Precision: Henry M. Leland*. Detroit: Wayne State University Press, 1966.

Lewis, Alfred Allan. *Ladies and Not-So-Gentle Women*. New York: Penguin Books, 2001.

Litvin, Faydor L. *Development of Gear Technology and Theory of Gearing* (NASA Reference Publication 1406). Cleveland: NASA Lewis Research Center, 1998.

Lytton, Sir Edward Bulwer. *Leila: or the Siege of Granada*. Paris: A. and W. Galignani, 1838.

May, Arthur J. *A History of the University of Rochester 1850–1962*. Rochester: University of Rochester, 1977.

McKelvey, Blake. *Rochester, The Flower City: 1855–1890*. Cambridge: Harvard University Press, 1949.

———. *Rochester, The Quest for Quality: 1890–1925*. Cambridge: Harvard University Press, 1956.

McNamara, Robert F. *The Diocese of Rochester in America: 1863–1993*. 2nd ed. Rochester: Roman Catholic Diocese, 1998.

National Housing Association. *Housing Problems in America: Proceedings of the Eighth National Conference on Housing.* New York: N.H.A., 1920.

Oldenziel, Ruth. *Making Technology Masculine: Men, Women and Modern Machines in America 1870–1945.* Amsterdam: Amsterdam University Press, 1999.

Pratt & Whitney. *Accuracy for Seventy Years 1860–1930.* Hartford, Conn.: Pratt & Whitney Co., 1930.

Quinn, Susan. *Marie Curie: A Life.* New York: Simon & Schuster, 1995.

Remington, Carolyn Lyon. *Of Double Interest: Because the Author Is a Twin: An Autobiography.* Rochester, N.Y.: C. L. Remington, 1986.

———. *Vibrant Silence.* Rochester: Lawyer's Cooperative Publishing Co., 1965

Rochester Engineering Society. *A Century of Engineering in Rochester 1897–1997.* Rochester: 1997.

Roe, Joseph Wickham. *Tool Builders: The Men who Created Machine Tools.* New Haven: Yale University Press, 1916.

Rosenberger, Jesse Leonard. *Rochester: The Making of a University.* Rochester: University of Rochester, 1927.

Ross, Claire. "Kate Gleason of Rochester: America's Pioneer Woman Machinist." *Pictorial Review,* September 1919, 29.

Rundschauer, "Just Looking Around" (obituary of Kate Gleason), *Cornell Alumni News* 35, no. 13 (January 19, 1933): 178.

Sherr, Lynn. *Failure Is Impossible: Susan B. Anthony in Her Own Words.* New York: Random House, 1995

Sinclair, Bruce. *A Centennial History of the American Society of Mechanical Engineers: 1880–1980.* Toronto: University of Toronto Press, 1980.

Slaton, Amy E. *Reinforced Concrete and the Modernization of American Building, 1900–1930.* Baltimore: Johns Hopkins University Press, 2001.

Sponholtz, Shirley. "Goldie Prame's 1914 Northway Trailer." *Old Time Trucks,* February/March 2004, 24–27.

Woodbury, Robert S. *History of the Gear-Cutting Machine.* Cambridge: M.I.T. Press, 1958.

Yost, Edna. *Frank and Lillian Gilbreth: Partners for Life.* New Brunswick: Rutgers University Press, 1949.

Index

Colophon

Typeset in Adobe Garamond Premier Pro.

Printed on Nature's Natural 30% post-consumer recycled paper and Huron Gloss 10% post-consumer recycled paper.

Printed by Thomson-Shore, a member of the Green Press Initiative.